4 遗迹，集体纪念碑

- 004 遗迹，集体纪念碑 _ Nelson Mota
- 008 DOMunder地下博物馆 _ JDdVarchitects
- 014 巨石堆砌的空间转换器 _ Estudi d'arquitectura Toni Gironès
- 026 圣莫里斯修道院考古遗址的覆层 _ Savioz Fabrizzi Architectes

34 另辟蹊径的文化建筑

- 034 另辟蹊径的文化建筑 _ Douglas Murphy
- 040 汇流博物馆 _ Coop Himmelb(l)au
- 058 蒙斯国际会展中心 _ Studio Libeskind + H2a Architecte & Associés
- 070 乌特勒支音乐宫 _ architectuurstudio HH + NL Architects
    + Jo Coenen Architects & Urbanists + Thijs Asselbergs
- 084 维加斯·阿尔塔斯会议中心和礼堂 _ Pancorbo + de Villar + Chacón + Martín Robles

104 办公室景观

- 104 办公室景观 _ Heidi Saarinen
- 110 空中食宿公司都柏林办公室 _ Heneghan Peng Architects
- 118 Sound Cloud总部办公室 _ KINZO
- 124 Kashiwanoha开放创意实验室 _ Naruse Inokuma Architects
- 130 Crosswater办公室 _ Threefold Architects
- 138 马德里马塔德罗文化工厂 _ Office for Strategic Spaces
- 148 IBC创意工厂 _ Schmidt Hammer Lassen Architects
- 154 Lowe Campbell Ewald总部 _ Neumann/Smith Architecture
- 164 1305工作室 _ 1305 Studio
- 172 CDLE办公室 _ R-Zero Arquitectos

- 180 建筑师索引

建筑立场系列丛书 No. 53

Ruins as Collective Monuments

004 *Ruins as Collective Monuments_Nelson Mota*

008 DOMunder_JDdVarchitects

014 Space Transmitter of the Mound_Estudi d'arquitectura Toni Gironès

026 Coverage of Archaeological Ruins of the Abbey of St-Maurice_Savioz Fabrizzi Architectes

Alternatives Strategies for Culture Buildings

034 *Alternatives Strategies for Culture Buildings_Douglas Murphy*

040 Confluences Museum_Coop Himmelb(l)au

058 Mons International Congress Xperience_Studio Libeskind + H2a Architecte & Associés

070 Utrecht Music Palace_architectuurstudio HH + NL Architects
    + Jo Coenen Architects & Urbanists + Thijs Asselbergs

084 Vegas Altas Congress Center and Auditorium_Pancorbo + de Villar + Chacón + Martín Robles

The Changing Landscape of the Office Interior

104 *The Changing Landscape of the Office Interior_Heidi Saarinen*

110 Airbnb Dublin Office_Heneghan Peng Architects

118 Sound Cloud Headquarters_KINZO

124 Kashiwanoha Open Innovation Lab_Naruse Inokuma Architects

130 Crosswater's Office_Threefold Architects

138 Cultural Factory in Matadero Madrid_Office for Strategic Spaces

148 IBC Innovation Factory_Schmidt Hammer Lassen Architects

154 Lowe Campbell Ewald Headquarters_Neumann/Smith Architecture

164 1305 Studio Office_1305 Studio

172 CDLE Office_R-Zero Arquitectos

180 Index

遗迹，集体纪念碑

Ruins as Collective Monuments

　　遗迹是反映我们现在环境的镜子，让我们更全面地认识自己和身居的世界。它们是激活我们集体记忆的要素。然而，我们需要一个物质或精神框架来搭建与遗迹交流的桥梁。介质是我们与遗迹交流的必要条件，它可使我们之间保持临界距离。艺术和建筑就是介质。事实上，艺术作品中常不乏对遗迹的描绘（例如，18和19世纪的绘画作品中遗迹比比皆是）。当前，尽管算不上是最关键的特性，但也不影响其成为建筑设计中不可或缺的一部分。

　　1961年Paul Zucker写到对遗迹的艺术处理有三种美学态度。这三种态度都涉及过去与现在的辩证关系。有的趋于将历史浪漫化；有的试图铭刻历史；还有的仅仅是想重现历史。然而，Zucker指出，采用单一美学理念的情况很少见，通常这些美学理念会相互结合使用，形成"美学混合体"。

　　现在在那些受到现存遗迹深刻影响的建筑设计作品中都能够发现美学混合体的影子。在这种情况下，建筑要多大程度地将遗迹改造为设施，才能使之激发出我们对所处世界的理解？这个问题越来越挑战建筑协调过去与未来的尺度。与考古学家一样，建筑师必须能够知晓隐藏在表面下的不同层次，越过表象构建现实。运用这种能力，建筑师激活我们的集体记忆。

Ruins perform as mirrors of our current condition, from which we learn a great deal about ourselves, and the world we live in. They are powerful elements to activate our collective memory. However, we often need a physical or intellectual framework to mediate our interaction with them. We seldom interact with ruins without an intermediary device, one that establishes a critical distance between them and us. Art and Architecture are some of these intermediary devices. Indeed, ruins are often depicted in works of art (they were omnipresent in 18th and 19th centuries painting, for example), and are nowadays a recurrent component, if not the main feature, in architectural commissions.

In 1961, Paul Zucker wrote that there were three aesthetic attitudes regarding the artistic approach to ruins. They were all related with a dialectical relation between the present and the past. In some cases there was a tendency to romanticize the past; in other cases the driving force was documenting the past; finally there were those interested in simply reviving the past. However, as Zucker points out, there are few cases in which one can find a pure aesthetic attitude. Rather, they are usually combined, creating "aesthetic hybrids".

It is this aesthetic hybridity that we can find today in architectural operations that are heavily influenced by the presence of ruins. In this context, thus, to what extent can architecture transform ruins into devices to activate our understanding of the world we live in? This is a question that increasingly challenges the architecture discipline in asserting its mediating role between the past and the future. Like an archaeologist, to go beyond the shallow layer that configures our reality, the architect has to understand the different strata hidden under the surface. Using this capacity, architects activate our collective memory.

DOMunder地下博物馆_DOMunder/JDdVarchitecten
巨石堆砌的空间转换器_Space Transmitter of the Mound/Estudi d'arquitectura Toni Gironès
圣莫里斯修道院考古遗址的覆层_Coverage of Archaeological Ruins of the Abbey of St-Maurice/Savioz Fabrizzi Architectes

遗迹，集体纪念碑_Ruins as Collective Monuments/Nelson Mota

在1961年发表的论文《遗迹，美学混合体》中，Paul Zucker强调欧洲现代艺术史中遗迹对美学理念发展的重要性。[1]然而，遗迹的重要性不仅仅局限于美学领域。事实上，正如美国地理学家J.B. Jackson在1980年发表的《遗迹的必要性》一文中说的，遗迹是联系过去的媒介，这比文物的美学价值和美学意义还重要。[2] Jackson声称，我们与历史的联系不是政治上的，而是民间乡土的过往。毕竟，我们真正珍视的是那些没有记载确切年代的日常生活琐碎。

在Jackson看来，遗迹是集体的纪念碑，因其能够唤起回忆。但是，和其他纪念品不同，作为集体纪念碑，遗迹与不同的历史时期相联系，"不是历史书上描述的历史，而是民间历史，是黄金时代，没有日期，没有姓名，满载着对以往生活的感知，是一部日常生活的编年史。"[3]

前面讲到的项目涉及的三种建筑方法将遗迹当作集体纪念碑。这三种方法可以呈现旧貌，让我们一睹过去的日常生活。这些项目的共同目的是将历史片段带到当下。在Paul Zucker所说的"美学理念"方面，它们之间进行了区分。

铭刻历史

DOMunder地下博物馆由JDdVarchitecten设计，位于荷兰乌特勒支市。它挖掘出了部分神秘的Domplein主教广场，以展现拥有2000多年历史的遗迹。1674年，乌特勒支Domkerk哥特式教堂的中殿被损毁，留下的空地便是现在的主教广场。飓风席卷之后，这座昔日荷兰最大的教堂只剩下高塔（主教塔）、圣坛和耳堂。四百多年来，中殿一直存在于想象中。现在，DOMunder地下博物馆使大教堂隐藏的部分重见天日。这项考古尝试展示了罗马城堡的遗迹，使更久远的文物重获生机。JDdVarchitecten的规划设计具有两面性。一方面，它的构造十分隐蔽，在广场方向很难发现，入口的标志仅仅是一段考顿钢条栅栏，这些钢条栅栏隐藏在广场地砖中间，与地面浑然一体，因而与广场空间布局之间的交界面达到最小。而在地下，效果却大相径庭。参观者从广场入口进入到地下，沿着步

In his essay "Ruins. An Aesthetic Hybrid", published in 1961, Paul Zucker stressed the importance of ruins for the development of aesthetic attitudes in European modern art history.[1] The importance of ruins, however, goes beyond the world of aesthetics. In fact, as the American geographer J.B. Jackson put it in his 1980 essay "The Necessity for Ruins", the ruins are a vehicle to create an association with the past, which goes beyond the artefact's aesthetical value or interest.[2] Jackson contended that in our relation with the past "the association seems to be not with our politically historical past, but with a kind of vernacular past". Indeed, according to Jackson "what we cherish are mementos of a bygone daily existence without a definite date."

For Jackson the ruin was thus a collective monument for its power to remind. However, differently from monumental art, the ruin as a collective monument creates a link with a different past, "not the past history books describe, but a vernacular past, a golden age where there are no dates or names, simply a sense of the way it used to be, and a history as the chronicle of everyday existence."[3]

In the projects featured ahead we can thus recognize three architectural approaches in which ruins perform as collective monuments. We can indeed observe three schemes that attempt to show us how it used to be, to give us glimpses of a bygone everyday life. While these projects share a common drive to exhibit and re-configure fragments of the past confronting them with our present circumstance, they can nevertheless be differentiated as to their "aesthetic attitude" as Paul Zucker would put it.

Documenting the Past

The project "DOMunder", designed by JDdVarchitecten for the Dutch city of Utrecht, excavates a part of the enigmatic Domplein, the Cathedral's square, to reveal remnants of twenty centuries of history. The Domplein is nothing but the void created by the destruction, in 1674, of the central nave of Utrecht's gothic cathedral, the Domkerk. After being wiped out by a tornado, the only remnants of the once largest cathedral of the Netherlands were the tower (the Domtoren), the choir and the transept. For more than four centuries, the presence of the absent nave was only virtual or ephemeral. Now, with the "DOMunder" project some hidden parts of the cathedral were brought to existence. Furthermore, this archaeological endeavour gave life to findings that go even further in time, unveiling remnants of the Roman forts. The scheme designed by JDdVarchitecten is somewhat ambivalent. On one hand they devised a structure that is barely perceptible from the square, signalled only by an access hatch made out of corten steel elements that can be integrated into the ground in such a way as to camouflage them with the square's brick pavement. They thus create a minimal interference with the spatial configuration of the square. Under the ground, the effect is di-

道,穿梭于古迹之间,不必担心会影响到它们。事实上,单体的考顿钢架在聚集了众多历史层面的空间里显得非常宏伟,DOMunder地下博物馆的目的很明确,铭刻历史,让参观者在"考顿钢之手"的指引下,探访古迹。

将历史浪漫化

和乌特勒支主教广场的DOMunder地下博物馆低调的表现形式相反,Estudi d'arquitectura Toni Gironès设计的巨石堆砌的空间转换器在西班牙小村庄Seró的景观内颇为引人注目。尽管建筑很显眼,但是它还是小心地融入当地地貌,巧妙地与周围建筑和自然景观融为一体。新建筑的主要目的是建造一座博物馆,展示和保护距今已有近5000年历史的、最近才被挖掘的墓葬和巨石结构,因为这是修建该建筑的主要原因,因此其功能和用途都有所拓展,并且该建筑成为小型文化设施和社交中心,包括多功能室和空间,以展示和出售当地社区制造的产品。建筑设计对材料的使用独具匠心,就地取材,与野兽派美学一脉相承。钢筋混凝土和陶瓷材料比比皆是,与酷似在建大楼外部使用的钢条的钢铁元素交织在一起。Toni Gironès使用了一系列的材料(粘土、铁、表层土和花岗岩石子),打造多重触觉体验。此外,房间的空间结构设计营造出一种漫步式建筑,不断挑战参观者的感官。通过考古遗迹、砖石混凝土墙体和印有每一位参观者足迹的粘土步道的完美结合,巨石堆砌的空间转换器建筑将短暂与永恒融为一体。

遗迹激发了大胆的设计主题,而Toni Gironès则重新定义了展示遗迹的浪漫方法。就像一个浪漫派画家,通过营造枯石与生长的植被的色彩反差,让光、影、细微变化的色彩有机会形成精彩的互动。[4]

重现历史

从考古和美学的角度来看,相对遗址进行了浪漫化处理,我们发现对其进行美学方面的处理则趋向于重现历史,激发空间想象,强调建筑和谐,比如空间和体积的关系,如Savioz Fabrizzi建筑师事务所设计的圣莫里斯修道院考古遗址的覆层。

圣莫里斯修道院有1500年的历史,其间,滑坡和山上的坠石不断地破坏这些建筑。然而修道院遗址改建工程并未将遗迹遮掩起来,修道

verse, though. From the square's access point the visitor goes underground and follows a pathway and moves between the archaeological findings not being afraid of competing with these. In fact, the corten steel monocoque is an imposing element in a space already populated with so many historical layers. The DOMunder shows a clear intention to document the past taking the visitor in an archaeological expedition guided by a "hand of corten steel".

Romanticizing the Past

As opposed to the inconspicuous presence of the DOMunder in Utrecht's Domplein, Estudi d'arquitectura Toni Gironès's project for the "Space Transmitter of the Mound" creates a noticeable element in the landscape of the small Spanish village of Seró. Although its presence is evident, it is nevertheless carefully integrated on the site's topography and cleverly articulated with the surrounding elements, both built and natural. The main purpose of the new building is creating a museological space to show and protect a tomb and megalithic structures with almost five millennia that were discovered recently. While this was the driving force to commission the building, its function and purpose were expanded and it became a small cultural building and a social center including multipurpose rooms and spaces to display and sell products made by the local community. The design of the building shows a cunning use of materials that bears affinities with the aesthetics of brutalism in its unhindered exhibition of materials "as found". The use of reinforced concrete and ceramic is pervasive, intertwined with steel elements that look like reinforcement bars of an unfinished structure. In the building's pavements Toni Gironès uses a sequence of materials (clay, iron, topsoil, and granite gravel) to generate multiple tactile experiences. Furthermore, the spatial configuration of the rooms was designed in such a way as to create a promenade architectue in which the senses of the visitor are constantly challenged. The building for the Space Transmitter of the Mound combines transiency with timelessness, integrating seamlessly the archaeological ruins with brick and concrete walls and clay powder pavements that record the footprints of the visitors.

Toni Gironès reconceptualises a romantic approach to the exhibition of ruins that stimulates motifs of design bravura and, like a romantic painter, creates "opportunities for the scintillating interplay of light and shadow, of nuances of color, provided for by the interesting contrasts between the tonal values of withered stones and growing vegetation."[4]

Reviving the Past

In contrast with the romantic approach to ruins we can find aesthetic approaches driven to revive the past and stimulate our spatial imagination, emphasizing architectural harmonies such as the relation between space and volume, from an archaeological or

院与悬崖相对,而悬崖上的巨石是摧毁修道院的原因,其作为该考古场地内的主角之一,与遗址争辉。这一效果是通过一个从悬崖边悬拉出的与修道院等高的屋顶来实现的,这个屋顶对遗址起到保护的作用。这个设计使新结构成为修道院的一部分。地面与遗址之间的交界面所产生的影响微乎其微,只有薄薄的木板小路蜿蜒穿过考古遗址,并与围合空间的雅致的金属墙相连。半透明悬空的屋顶进一步激活了那些"建造"遗址的事件的记忆。屋顶上仍堆放着170吨的岩石,以保护结构,抵御大风,这也寓意着这些岩石是造成损毁的罪魁祸首。通过遮挡修道院遗址的石头屋顶,设计者巧妙地将功能要求和强烈的象征意义相结合,他们说:"它暗示场地内这一持续的危险暴露在人们眼前。"历史就这样以含蓄而又有力的方式被复原和重构。

美学混合体

所有这些作品中都有精心设计的穿越遗迹的步道,而这些遗迹组成了整个项目的材料和本体。在这种双重属性中,遗迹是界定嵌入空间布局的关键因素,同时,遗迹还是展览的主体。模糊化的处理激发参观者来理解遗迹的不同历史层面。

事实上,上述项目展示了将遗迹作为历史文物进行处理的三种不同的美学理念。

DOMunder地下博物馆主要记录了过去的时光,巨石堆砌的空间转换器创造了将历史浪漫化的方法,圣莫里斯修道院考古遗址的覆层优雅地再现了历史,然而,没有一个项目使用的是单一的或者纯粹的美学方法。三个项目都混杂了大量的、不同的美学观点。这些项目是美学混合体,用料考究,空间布局精巧,激发了我们的集体记忆,并且帮助我们理解所处的世界。

even aesthetical viewpoint. This is the case of the project for the Coverage of Archaeological Ruins of the Abbey of St-Maurice, designed by Savioz Fabrizzi Architectes.

This project bluntly exposes the ruins of the parts of the 1500 years old Abbey of St-Maurice destroyed by landslides and rock falls over its lifetime. The cause of this destruction, the steep cliff against which the Abbey was built, competes with the ruins as one of the main protagonists in the archaeological site. This effect is achieved by a design decision to protect the ruins with a roof suspended from the cliff at the same height of the Abbey. This design makes the new structure part and parcel of the volume of the Abbey. At ground level, the interference with the ruins is reduced to a bare minimum, with only a slender wooden deck that meanders through the archaeological site and connects with the discreet metallic wall that encloses the space. The translucent suspended roof further activates the memory of the events that "created" the ruins. The 170 tons of rocks that were put on its top to protect the structure against windblasts becomes an allegoric representation of the agent of destruction. The designers playfully combine functional requirements with the strong symbolism conveyed by the "stone-roof" that covers the archaeological ruins of the Abbey. "It expresses", they argue, "the ever lasting hazard the site has been exposed to." The past is thus revived and reconstructed in a subtle yet powerful manner.

Aesthetic Hybrids

In all these works there is a carefully planned promenade through the ruins, which are both the material and the object of the project. In this dual capacity, they are crucial elements to define the spatial configuration of the intervention and, at the same time, they stand as the main pieces displayed in the exhibition. Sparked by this ambiguity, the observer is challenged to make sense of the different historical layers.

Indeed, the projects discussed above show three different aesthetic attitudes in dealing with the ruins as artifacts.

The DOMunder is chiefly engaged with documenting a bygone time, the Space Transmitter of the Mound creates a new way of romanticizing history, and the Coverage of Archaeological Ruins of the Abbey of St. Maurice gracefully revives the past. None of them is, however, pure or monolithic in its aesthetical approach. There is a great deal of contamination in all of them, mixing and negotiating different aesthetic attitudes. These projects are aesthetic hybrids that show a careful use of materials and spatial configurations to activate our collective memory and to help us in making sense of the world in which we live in. Nelson Mota

1. Paul Zucker, "Ruins. An Aesthetic Hybrid", *The Journal of Aesthetics and Art Criticism* 20, no.2 (December 1961), p.119~130.
2. John Brinckerhoff Jackson, "The Necessity for Ruins", *The Necessity for Ruins, and Other Topics* (Amherst: The University of Massachusetts Press, 1980), p.89~102.
3. Ibid., p.94~95, Emphasis original.
4. Paul Zucker, "Ruins. An Aesthetic Hybrid", p.120.

DOMunder地下博物馆

JDdVarchitecten

位于荷兰乌特勒支历史中心的DOMunder地下博物馆是一个新建的考古体验中心。其设计意图是通过重新挖掘考古遗迹来建造地下公共区域，使主教广场的隐秘历史重新展示在公众面前。

早期文物检测挖掘工作的文献材料都是以岩石层或栩栩如生的图画的形式保存下来的，基于对这类材料的广泛且综合的研究，一个虚拟的三维重构建筑在原有结构的上面展现出来。

主教广场方向的地下博物馆入口是由一系列的考顿钢条组成的，钢条折叠起来形成封闭的表面和开口。一段单独的楼梯井通往地下空间。楼梯演变成一段小道，而这段小道又演变成通往更低层空间的台阶。

由于哥特式柱子的柱石扎根于约5.5m深的地表之下，因此在挖掘过程中，人们随处可见不同时期遗留下来的元素和残骸，且其地面轮廓和地基变得几乎清晰可见。可持续使用的板材建成的一段步行通道，其材料特质正好适应湿度在75%以上的气候条件。

这段通道一方面可以指引参观者沿着一段多媒体墙壁行走，另一方面，参观者还可以通过一段真实可触摸的考古体验来感受其不可磨灭的荣耀历史。考顿钢衬上布满了数控技术打孔形成的小孔，成百上千，大小不一。通过背景光的照射，光线汇聚形成一幅点彩画，来展现历史的图景。这种设置与投影相结合，产生了极好的效果，为DOMunder博物馆的地下探索之旅增添了一种动态的、刺激的空间特性。

来到DOMunder地下博物馆的参观者使用专门开发的交互式手电（基于红外线和3D技术制成），下至地下探寻考古遗迹，听它们讲述主教广场两千年的历史传奇。

参观者带着一个手电筒来到地下，并且进入了考古学家的角色，成

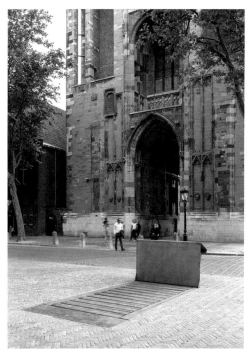

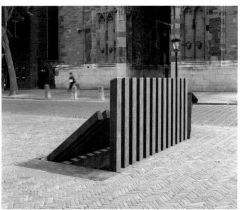

为一个时间旅者。他们通过这种方式来探索这个广场的神奇历史故事。另一个值得一提的视听特色是"可怕的风暴",这是参观者身处地下期间不得不承受的不适之感。仿佛1674年那场损毁中殿的风暴再次侵袭了广场。通过最先进的电脑动画、照明和音效技术,参观者重临了这次历史性的风暴。

将该项目由设想转化为现实,不仅仅要依靠现代科技,还要依靠传统工艺的知识与技巧。比如,利用光点成像的绘图技术实际上已经过时了。然而,要想建造拥有独特外形的衬层和过道一侧的栏杆,仅仅依靠现代科技是不足以将其转化为独特的三维形式的。只有经验丰富的专业人士才能够运用技艺将板材制作成复杂的外形,使其看起来是将材料从二维形态弯折成三维形态的。

DOMunder

DOMunder is a new archaeological experience center in the historical heart of Utrecht. The purpose of DOMunder is to make the concealed history of the Domplein accessible and perceptible, by creating subterranean public areas by way of re-excavating archaeological remains.

On the basis of an extensive and comprehensive research of documentation materials of the earlier test excavation(s) in the form of stratigraphy and photographs amongst other things, a virtual 3D reconstruction was made of the original mass that was present. To enable access from the Domplein, an entrance was designed which consists of a series of Corten steel fingers which, having been folded together, form a closed surface and hatch. The underground space is accessible by way of a single stairwell. The stairwell becomes a pathway, and the path in turn is formed into steps to an even lower pathway.

Since the Gothic pillars have their mainstays on solid ground at more or less 5.5m below the surface level, the ground profiles and foundations of all the layers of time became visible during the excavations, with different elements and remains of different periods at every place. A walking route was developed of sustainable sheeting material well suited for climatic conditions of 75% or more humidity.

On one side this pathway guides the visitor along a multimedia wall, and on the other side along the directly tangible archaeology in all its undiluted glory. This corten lining is perforated with a numerically controlled range of hundreds of thousands of dif-

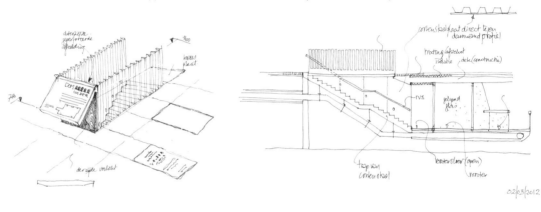

项目名称: DOMunder
地点: Domplein, 3512 Utrecht, the Netherlands
建筑师: JDdVarchitecten / 体验设计: Tinker imagineers
施工承包商: Aannemersbedrijf Van Zoelen
地面施工单位: Theo Pouw Groep / 施工单位: ABT
设施安装和气候控制: LBP
施工管理: Aestate
有效楼层面积: 350m²
设计时间: 2005—2014 / 施工时间: 2012—2014
摄影师:
©Mike Bink (courtesy of the architect) -p.11, p.13 bottom
©Oliver Schuh (courtesy of the architect) -p.8, p.10, p.12, p.13 top

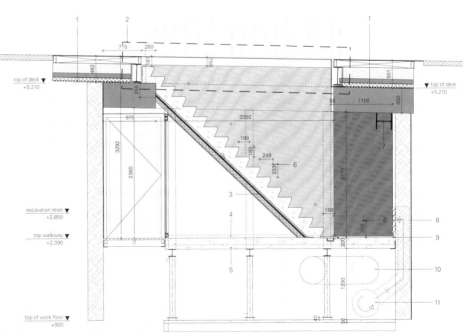

a-a' 剖面图 section a-a'

1. build-up of roofing structure in conformity with Hollanddak regulations/standards **2.** elaboration of access hatch to be coordinated with Nering Bögel **3.** space for drainage 30mm/insulation t.b.a./bitumen/drainage gutter **4.** option: installation of pressure-resistant insulation with hard topping on concrete floor **5.** floor: concrete on ComFlor 75 **6.** lighting in corten steel **7.** temporary brace **8.** profiled corten steel for positive pressure air outlet **9.** steel strip for affixing bitumen **10.** return ø500 **11.** positive pressure ø500

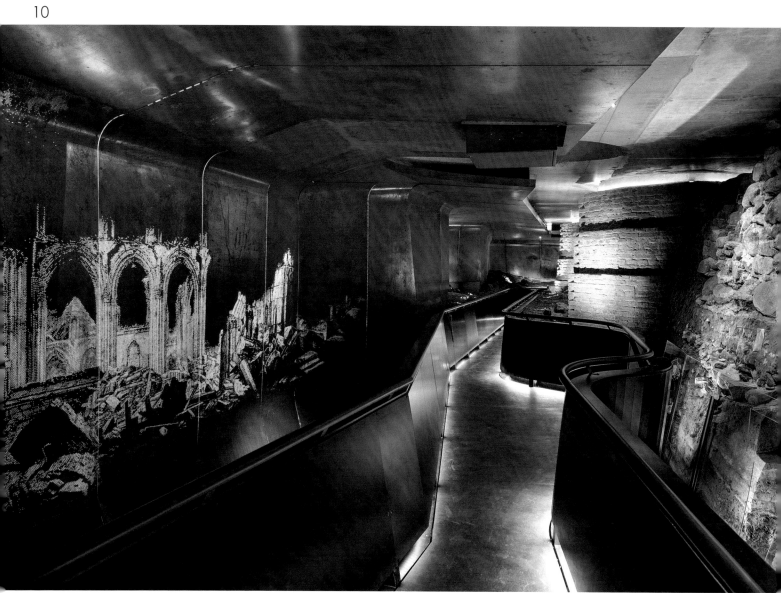

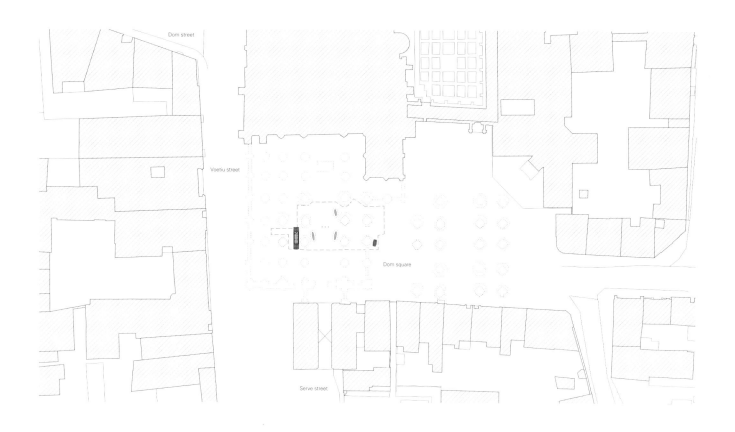

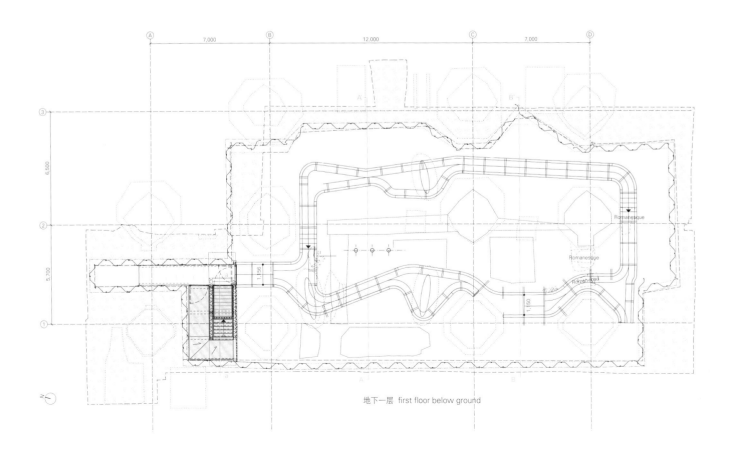

地下一层 first floor below ground

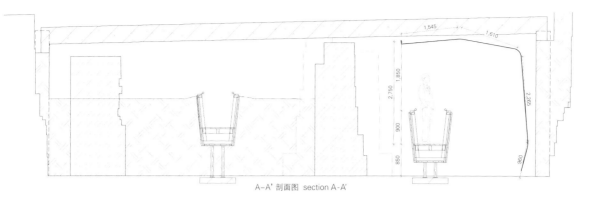

A-A' 剖面图 section A-A'

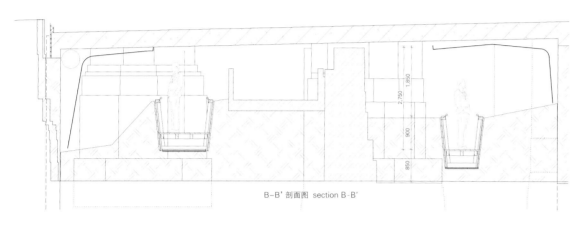

B-B' 剖面图 section B-B'

ferent-sized little holes. By backlighting these, together, they form pointillist images of historical scenes. This provides a superb effect in combination with projections, thus giving the underground discovery DOMunder a dynamically exciting spatial character.

DOMunder's visitors use a specially developed interactive flashlight (based on infrared and 3D technology) while descending in the dark in search of archaeological remains, which tell the stories of 2000 years of history at the Domplein. By descending underground with a flashlight, visitors assume the role of the archaeologist, a time-traveller. In this way they uncover in their search the unbelievable history of the square.

A second audiovisual highlight is "the terrible tempest" that overwhelms visitors during their stay under the ground. It's as if the storm that destroyed the nave in 1674 rages once again over the square. By making use of the state-of-the-art computer animations, lighting and sound effects, the visitor relives this historical storm.

This project has been made possible not only by virtue of modern technology, but especially because of the knowledge and expertise of traditional craftsmanship. The technology of the portrayal, for example, of an image by means of an accumulation of light points is actually decidedly archaic. However, in order to achieve the uniquely shaped lining or a balustrade of the walkway, this method was inadequate to effectively turn this into a unique 3D form. Only professionals who are highly experienced with the processing of sheeting material into complicated shapes appear to be able to combine curved 2D planes into unique 3D forms.

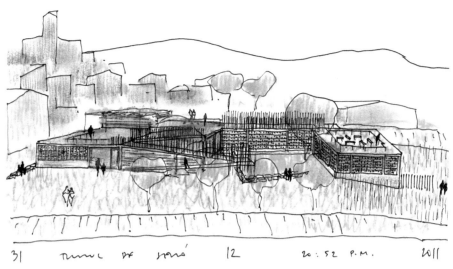

巨石堆砌的空间转换器

Estudi d'arquitectura Toni Gironès

2007年1月，在对位于Segarra-Garrigues的一条二次灌溉管道进行施工的过程中，人们意外地在西班牙Seró发现了一处公元前2800年的考古遗址。该发现的最重要的特点是除了采用巨大的砂岩板材之外，其表面更是精妙地雕刻了带有几何图案的浅浮雕装饰。

新建筑建于广场附近的两座废弃的果园内，这里将成为一处为当地居民提供多重空间的多功能文化中心。建筑材料就地选取，建筑结构巧妙地连接了广场与果园之间的垂直落差。

一系列柔软灵动的波纹钢围栏围成的缓坡不仅提供了不同的路径选择，还覆盖了公共空间的不同地势。一个由粘土和泥土建成的平台位于广场与"前比利牛斯山脉"构成的地平线之间，成为观望四个石柱的一处观景点，参观者可以在这里一览这处考古遗址的全貌。

最后，参观者沿一条坡度小到无法察觉的螺旋路径进入到石柱围合的房间内，周围环绕着陶瓷片，它们过滤了阳光和空气、乡村的气息和雾。光线强度被削弱，陶瓷路面瓦解，头顶的光线无声地汇聚在石柱的边缘表面。时间在细细地凝视中放慢了脚步，粘土粉铺就的水平地面上印下了每一位参观者的足迹，人们走着走着便不知不觉地从另一侧走了出去，任意两条道路都没有相交点。逐渐增强的光线和嘈杂的声音一路伴随着参观者，直到一处麦田形成的地平线忽然闯入视线，才将人们带回到寻常的农业景观中。

Space Transmitter of the Mound

In the month of January of 2007, the construction work on a secondary irrigation line in Segarra-Garrigues, caused the unexpected discovery of archaeological ruins in Seró, dating back to 2,800 BC. The most important aspect of this discovery besides the megalithic nature of the sandstone slabs was the accuracy of the sculpted geometrical low relief decorations.

A cultural center was to be built on the terrain composed of the two unused orchards adjacent to the square, which would offer versatile uses and spaces. The projected construction was to be made with local materials and would be conceived in a way to bridge the vertical gap between the square and the orchards.

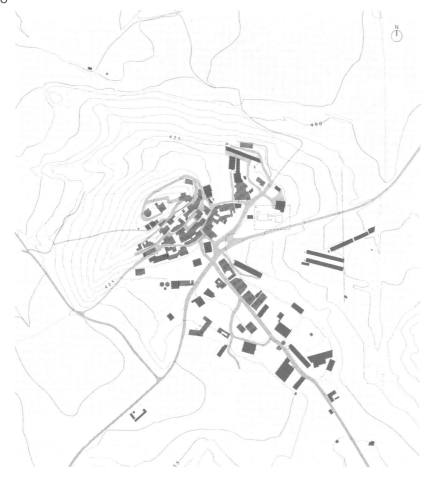
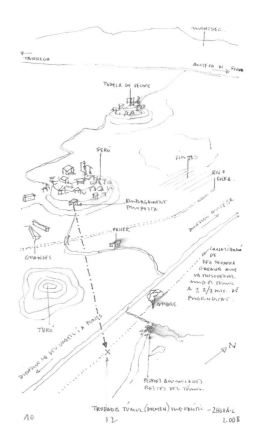
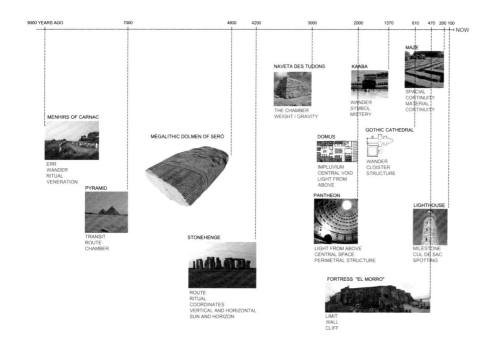
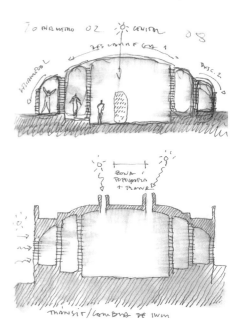

项目名称：Space Transmitter of the Mound
地点：Seró-Artesa De Segre, Lleida, Spain / 建筑师：Toni Gironès Saderra
项目团队：collaborator office _ Dani Rebugent, Technical architect _ Brufau I Cusó S.L.P.,
考古学家：Pep Castells I Joan López Arqueòlegs,
施工经理：Ignasi Gilabert, Furniture _ Fusteria Giribet S.L.
结构工程师：BOMA INPASA S.L.P., ESTUDI XV S.C.P.
工程师：Oriol Vidal Ingenieria S.L.P.
施工团队：Construcciones Orgèl.lia S.L., Construcciones Germans Gilabert S.L.
土木工程勘察：Instal.lacions Vilana S.L.
用地面积：3,615m² / 总建筑面积：503m²
造价：EUR 42/m²(exterior), EUR 390/m²(interior)
设计时间：2007 / 竣工时间：2007—2013
摄影师：©Aitor Estevez (courtesy of the architect)(except as noted)

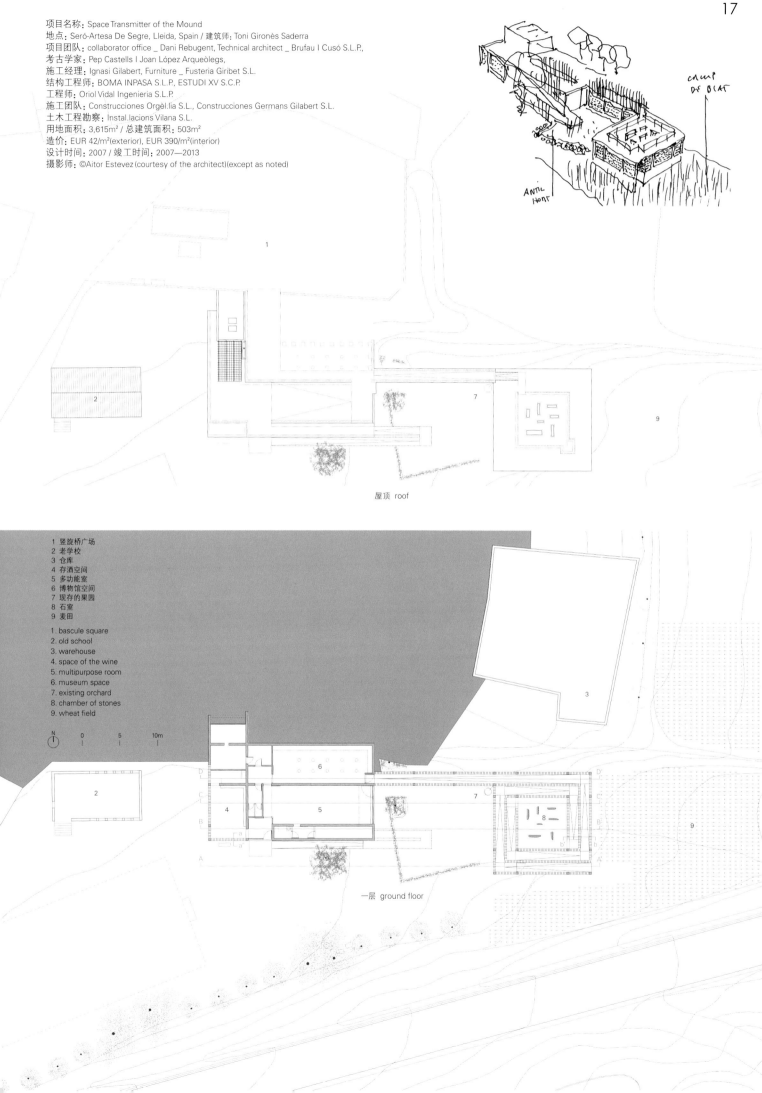

1. 竖旋桥广场
2. 老学校
3. 仓库
4. 存酒空间
5. 多功能室
6. 博物馆空间
7. 现存的果园
8. 石室
9. 麦田

1. bascule square
2. old school
3. warehouse
4. space of the wine
5. multipurpose room
6. museum space
7. existing orchard
8. chamber of stones
9. wheat field

屋顶 roof

一层 ground floor

It was projected to build a succession of mild ramps with soft and elastic limits made of corrugated steel that suggest different paths and cover different conditions of the public space. A clay and earth platform appears between the square and the "pre-Pyrenean" horizon, and this plane works as a viewpoint of the four steles from where people can observe the archaeological ruins. Finally, people start entering the room of the stele along a spiral tour with an almost imperceptible slope, surrounded by ceramic pieces that filter the light as well as the air, the smells of the countryside, and the fog. The light intensity weakens, the ceramic pavement disintegrates and the overhead light focuses the eye on the edged surface of each of the stele, in a silence. Time slows down in a space of accurate contemplation, in a horizontal plane of clay powder showing the footprint of each visitor, and quietly they start to walk out from the other side, with no possibility of the two paths ever crossing. Gradually intensified light and noise accompany people, until the horizon of a wheat field comes to find them and takes them back to the common agricultural landscape of the area.

Estudi d'arquitectura Toni Gironès

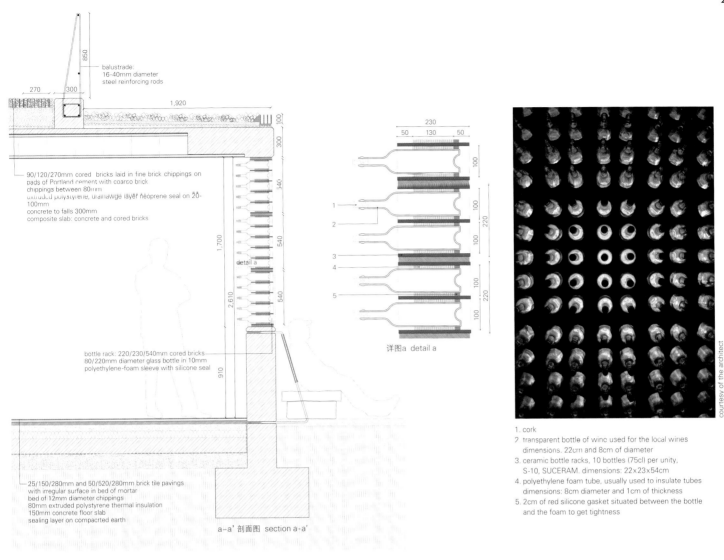

90/120/270mm cored bricks laid in fine brick chippings on pads of Portland cement with coarco brick chippings between 80mm
extruded polystyrene, drainage layer neoprene seal on 20-100mm
concrete to falls 300mm
composite slab: concrete and cored bricks

balustrade:
16-40mm diameter steel reinforcing rods

bottle rack: 220/230/540mm cored bricks
80/220mm diameter glass bottle in 10mm polyethylene-foam sleeve with silicone seal

25/150/280mm and 50/520/280mm brick tile pavings with irregular surface in bed of mortar
bed of 12mm diameter chippings
80mm extruded polystyrene thermal insulation
150mm concrete floor slab
sealing layer on compacted earth

a-a' 剖面图 section a-a'

详图a detail a

1. cork
2. transparent bottle of wine used for the local wines dimensions. 22cm and 8cm of diameter
3. ceramic bottle racks, 10 bottles (75cl) per unity, S-10, SUCERAM. dimensions: 22×23×54cm
4. polyethylene foam tube, usually used to insulate tubes dimensions: 8cm diameter and 1cm of thickness
5. 2cm of red silicone gasket situated between the bottle and the foam to get tightness

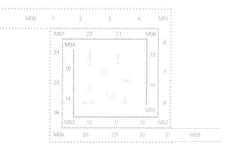

室内路线
interior route

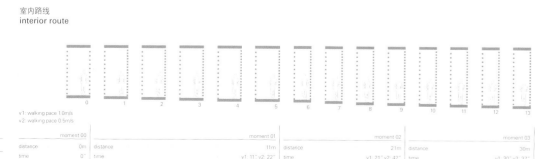

v1: walking pace 1.0m/s
v2: walking pace 0.5m/s

	moment 00	moment 01	moment 02	moment 03
distance	0m	11m	21m	30m
time	0"	v1: 11" v2: 22"	v1: 21" v2: 42"	v1: 30" v2: 37"

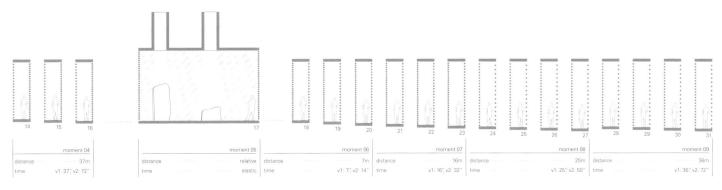

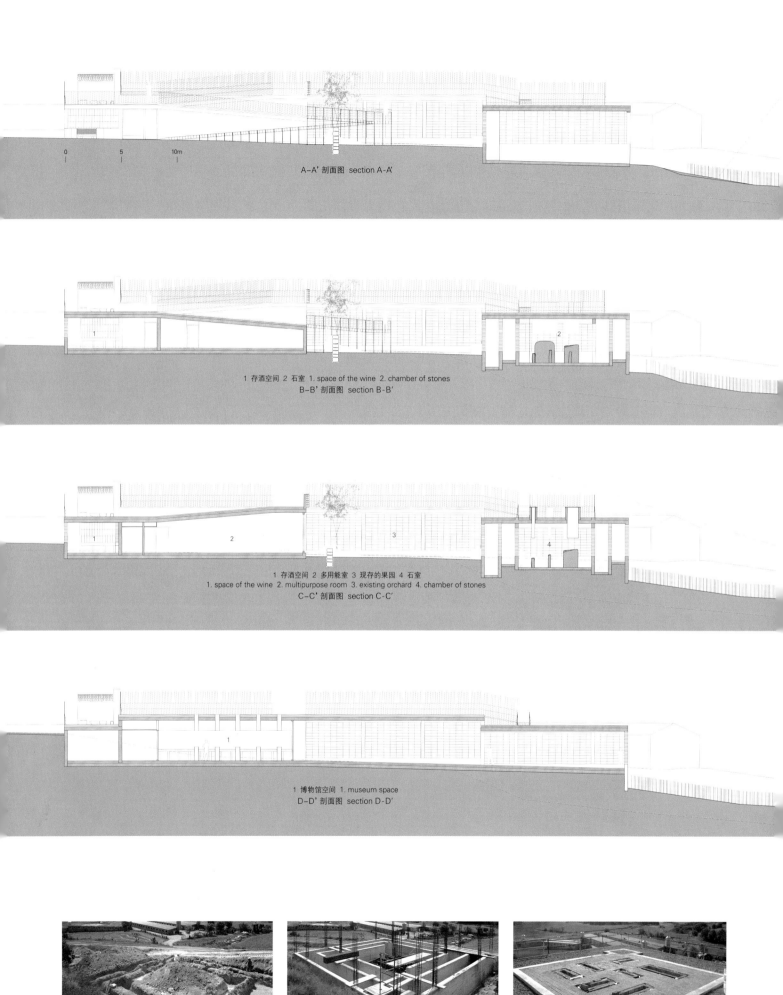

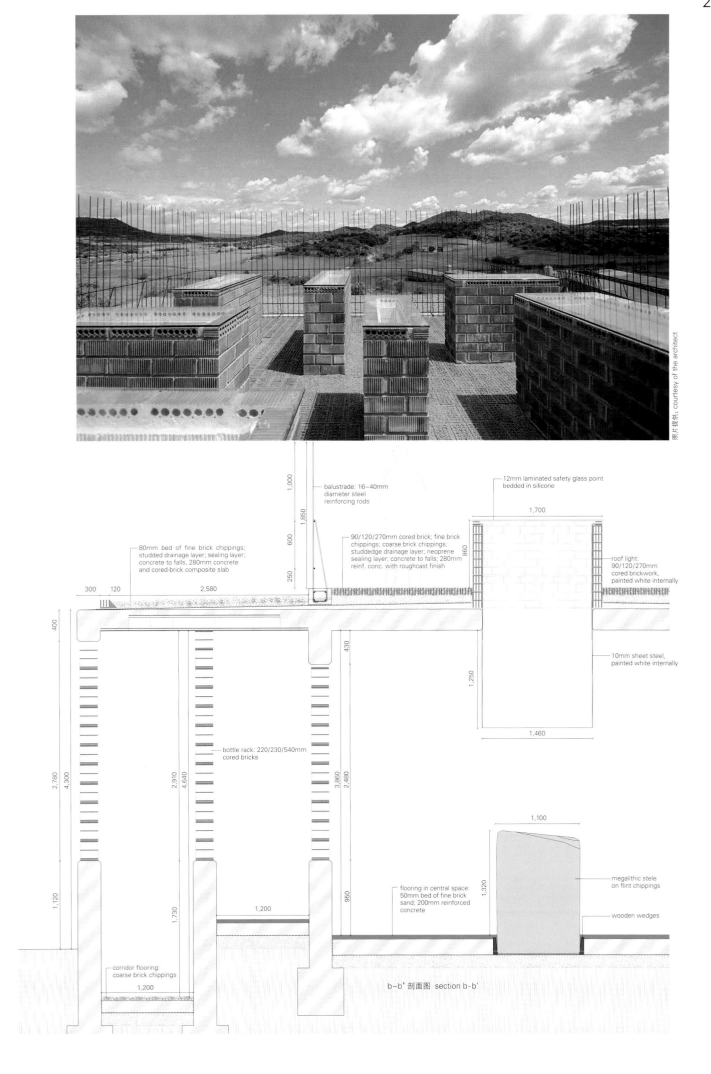

b-b' 剖面图 section b-b'

圣莫里斯修道院考古遗址的覆层

Savioz Fabrizzi Architectes

建于1500多年前的圣莫里斯修道院位于日内瓦和辛普朗之间的一段公路上,倚靠于一处悬崖之下。这个选址最初是出于悬崖提供的防御优势的考量。可惜千虑一失,圣莫里斯修道院最终却被从悬崖上跌落的巨石摧毁,可见当初防御要塞的构想是错误的。

纵观历史,落石是导致该修道院损毁的最主要原因。1611年,一块巨石跌落使建筑受到了严重的撼动,1942年,一块落石又砸毁了建筑的十字尖顶和入口中殿。历史上滑石事故不断,一次次在场地内留下了印记。

该项目有意将建筑所经历的特殊历史遭遇表现出来。设计师刻意在建筑上方悬拉的屋顶上放置了170吨石块,来表现修道院长久以来始终面临着的风险。然而此屋顶带给室内的反而是一种静谧到几近可以冥想的氛围。石材屋顶结构建于原有建筑之上,实现了外立面与悬崖表面的"对话"互动,并且作为光线的过滤器,为其下的内部空间提供均匀、稳定的照明。

Coverage of Archaelogical Ruins of the Abbey of Saint Maurice

The Abbey of Saint Maurice was built almost 1500 years ago. It is situated against a cliff in a section of the road between Geneva and the Simplon Pass. This particular site was likely to be chosen for its defensive position given by the cliff. However, the planned fortification proved a failure due to rock fragments falling from the cliff on to the building.

Throughout the history rock slides had an important influence causing several damages: In 1611 a huge rock fall caused a convulsion of the abbey, in 1942 a rock destroyed the cross-spire and

portal-nave of the building. Those slides have constantly been occurring and marked the site throughout history.

This project endeavors to demonstrate this particular treat of history. By suspending 170 tons of stones, it expresses the ever lasting hazard the site had been exposed to. The roof gives the interior a calm and almost contemplative ambience. It is built above the fundamental buildings in order to keep the dialogue between the facades and the cliff face. The "stone-roof" serves as a filter for light and produces an evenly spread and regular illumination.

Savioz Fabrizzi Architectes

项目名称：Coverage of Archaeological Ruins of the Abbey of Saint Maurice
地点：Saint Maurice, Switzerland
建筑师：Savioz Fabrizzi Architectes
土木工程师：Alpatec sa / 甲方：The Abbey of Saint Maurice
功能：Protection roof for the archaeological ruins
总建筑面积：1,400m² / 设计时间：2004 / 施工时间：2009—2010
摄影师：©Thomas Jantscher

32

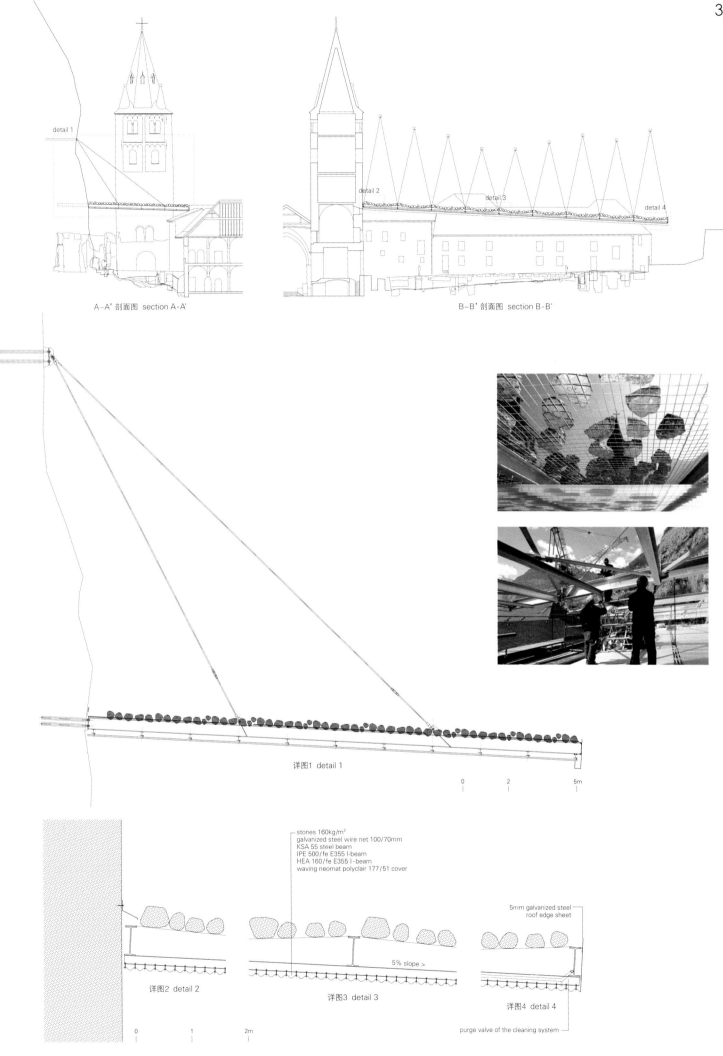

A-A' 剖面图 section A-A'

B-B' 剖面图 section B-B'

详图1 detail 1

0   2   5m

- stones 160kg/m²
- galvanized steel wire net 100/70mm
- KSA 55 steel beam
- IPE 500/fe E355 I-beam
- HEA 160/fe E355 I-beam
- waving neomat polyclair 177/51 cover

5mm galvanized steel roof edge sheet

5% slope >

详图2 detail 2

详图3 detail 3

详图4 detail 4

0   1   2m

purge valve of the cleaning system

另辟蹊径的文化建筑
Alternative Strategies for Culture Buildings

当代建筑的设计潮流仍是以那种大胆、令人兴奋、有视觉冲击感的建筑设计为推动力量。在重建和新建的城市中心,文化建筑的成本预算更多,而且还能够获得最显赫的名声。假如这种建筑形式方面的试验不再占据建筑设计的最前沿位置,那么建筑师还有其他方式来完成他们的使命吗?

文化建筑是一种形式极富表现力的建筑设计方法。它的功能通常展现为大型室内空间,如美术馆或是大礼堂,不需要与外部进行连接,允许建筑师对外围护结构的形式进行大量的处理。由于公众频繁地使用这些建筑,因此建筑更有可能被认为是公共空间的延伸,像公园和广场一样,这些开放的场景在周围的环境中显得卓然独立,创造了其边界与城市清晰隔开的景象。

但这种文化建筑类型也有很多变化,随着时间的推移,这种变化越来越多,建筑大师们不断地尝试和开发数字化试验,同时利用新型材料和新功能的设计方法也体现了建筑的前卫感。

The demand for bold, exciting, visually striking buildings continues to be the driving force behind a whole stream of contemporary architecture. The biggest budgets and the greatest fame are afforded to the creation of cultural buildings at the heart of remade or newly made cities. But if this kind of formal experimentation is no longer the cutting edge, are there other ways in which architects can approach these commissions?
The cultural building lends itself to formally expressive architectural approaches. Its functions are often large interiors such as galleries and auditoria that require disconnection from the outside, allowing for much formal manipulation of the external envelope. The buildings also, thanks to their intense public usage, encourage being considered as outgrowths of public spaces such as parks and plazas, and these open vistas often accentuate the singularity of the buildings, creating images whose boundaries are clearly separated off from the city.
But within this typology there is still variation, indeed more so than there has been for some time. The established names continue to practice and develop their digitally influenced experiments, while other approaches include functional and material innovations in the service of architectural boldness.

汇流博物馆_Confluences Museum / Coop Himmelb(l)au
蒙斯国际会展中心_Mons International Congress Xperience / Studio Libeskind + H2a Architecte & Associés
乌特勒支音乐宫_Utrecht Music Palace / architectuurstudio HH + NL Architects + Jo Coenen Architects & Urbanists + Thijs Asselbergs
维加斯·阿尔塔斯会议中心和礼堂_Vegas Altas Congress Center and Auditorium / Pancorbo + de Villar + Chacón + Martín Robles
另辟蹊径的文化建筑_Alternative Strategies for Culture Buildings / Douglas Murphy

21世纪的今天，在大型公共建筑的设计中，建筑师会做出什么样的选择呢？那些有着雄厚实力和资源的初级战略仍然保留着迄今为止已有的标志性建筑物的传统，而内在的文化功能则在全世界闻名的建筑师的妙笔下，被包裹在昂贵的、富丽堂皇的"盒子"内。这种设计非常适合正在进行的、大面积重建的区域，因为其周围的城市环境也几乎都在重建中。虽然这种设计理念最为盛行，但一些资源较少的、小规模的、以甲方的意愿为主的建筑也是可行的。

如果那些预算昂贵的、在全球都具有标志性的工业建筑设计行不通的话，有些不是很极端的设计方法也可择路而行，这些设计形式并非大胆前卫，而是巧妙地使用建筑材料，并且在规划和功能方面进行了创新。有许多年轻的设计师都可以传递这些理念。看看近期西欧建成的四座文化建筑，他们的建筑师有享誉全球的大师，有后起之秀，从这些建筑设计中我们可以看到一系列潜在的设计方法。

汇流博物馆位于法国里昂市的两条河流之间的一片狭长的陆地上，于近期开放，是一座关于科学和人类学的博物馆。它由蓝天组建筑事务所设计，极具雕塑感。博物馆位于显要的位置，人们站在城市周围的山谷斜坡上都能望见它，其设计充分地表现了奥地利建筑师的标志性用材——玻璃与钢架，它们构成的体量形成了复杂且变化莫测的形式。

该建筑包含一组"黑盒子"式的空间作为展厅，并肩负普通博物馆的功能——大礼堂、教育空间、零售店和餐饮中心。但是该建筑也尝试建造一处公共空间，使之成为半岛一端的新公园的一部分。为达成这一战略，设计需要提升建筑，将其置于基座上面，而基座下的某些建筑功能虽然被隐藏了，但是基座是完全开放的，行人可以从这座建筑底下直接穿过。

建筑师采用两个完全不同的体量来表现建筑。第一个体量为"水晶体"，是利用钢架和玻璃建成的主入口亭，它如同一个城市广场，建筑内

At this stage in the 21st century, what options are available for the construction of large public buildings? The primary strategy, for those with the power and resources, remains by now the long established tradition of the iconic edifice, where inward-facing cultural programme is wrapped in a very expensive decorative box by a world famous architect. This option is most suitable for areas that are undergoing extensive redevelopment, where the surrounding urban context is being almost entirely remade. But even though this may be the most prominent, there are other options available, ones that come into consideration for clients with less resources, or for commissions of smaller scale.

If the often massively overspent budgets of the global icon industry are not available, then less extreme architectural approaches are still possible, where formal boldness is exchanged for ingenious use of materials, or programmatic and functional innovations, and there are plenty of younger architects who can be commissioned to deliver these. By looking at four recent cultural buildings in Western Europe, designed by architects ranging from global stars to developing talents, a selection of these potential approaches can be seen in action.

The Confluences Museum is a science and anthropology museum that has recently opened on a small sliver of land between two rivers in Lyon, France. Designed by Coop Himmelb(l)au, it is a highly sculptural building, in a prominent location, visible from the valley slopes all around the city, and consists of the Austrian architects' trademarked steel and glass volumes arranged into complex and jarring forms.

The building contains a number of "black box" exhibition halls, as well as the standard museum functions: auditoria, educational spaces, retail and catering. But the building also attempts to create a public space, becoming a part of a new park that has been created at the end of the peninsula. The strategy for achieving this was to raise the building onto a plinth, under which certain functions are buried, but making this plinth almost completely open, so that pedestrians can pass directly under the building.

The form of the building itself is described by the architects in terms of two discrete volumes. The first is "The Crystal", which is the steel and glass entrance pavilion. This is described as being "like an urban square", and much of the circulation through the building is brought through here, around spiralling ramps and open staircases that combine with the angular structure to create a busy, invigorated space. To one side of this space is the "gravity

汇流博物馆，里昂，法国
Confluences Museum in Lyon, France

多条流线汇聚于此，围绕在螺旋形的斜坡和开放的楼梯的周围，斜坡和楼梯与棱角分明的结构结合在一起，形成了一处热闹的、富有活力的空间。这处空间的一侧设有一个"自流井"，给人以突然下降的视觉效果，其形状类似于爱因斯坦重力说里所描述的经典台球桌形象。这个自流井可以收集雨水，建筑本身在这里变成了展览。

建筑的另一个主要形式是"云朵"，这个结构内设有展厅，同时还兼具其他功能。这是一个钢架盒形结构，覆盖多种类型的三角形嵌板，嵌板的弯曲形式与建筑的形式相同，使游客有机会拾级而上，一直走到屋顶平台的开放空间。

该建筑历经了很长时间才建成，到2001年才完成设计竞标。其建筑设计沿用蓝天组建筑事务所一贯的风格——既有其早期作品中的角度并置的影子，又有其数字化设计作品中采用的半平滑形式。当形状试图变形或者重叠时，精心设计的、较难处理的嵌板就展现在眼前，其宛如蜥蜴骨骼一般的三角形结构凸显出来。这也证明了20世纪90年代试验性数字化时代所展现的功能适应性是完全有用的。在该建筑中，以及在盖里、艾森曼或其他建筑师设计的建筑中，数字化形式的发现过程始终与简洁的物理功能模型并驾齐驱（通常是无窗美术馆或礼堂的泡沫模型），由此形成外形的层叠性，从简单的内部空间，渐次过渡到复杂的、具有象征意义的建筑外围护结构，其唤起人们回忆的命名策略与建筑的整体形象紧紧相连。

比利时蒙斯国际会展中心与汇流博物馆的形式有着相似之处：前者在近期重建的城市中成为焦点，不仅与交通基础设施相连，而且还提供了项目书要求以外的公共空间。

该中心是由李博斯金工作室和H2a联合建筑事务所共同设计，虽然有着李博斯金的标志性形式设计手法和主题表现方式，但却避免了与过去设计的建筑相关的形式附属感。建筑最初的规划是三座大小不同的礼堂和一间开放的活动大厅，但是实际上建筑还设有一座大型屋顶花园以及一个观光平台，可以回望具有历史意义的蒙斯市中心和各种城市地标。

丹尼尔·李博金斯的设计方法是建造一些交织的"带状"结构，这

well", a sudden drop in the structure, shaped like the classic "billiard table" diagram of Einsteinian gravity, which will collect rainwater, the building becoming an exhibit in itself.

The other primary form is called "The Cloud", and this contains the exhibition halls and other programmes. This is a steel framed box, covered in vast series of triangulated panels that follow the twisted form of the building, with the opportunity for visitors to make their way out into the open air on platforms at the top.

This building has been a long time coming, with the competition originally being awarded in 2001. This is visible in Coop Himmelb(l)au's design, which is caught between the angular juxtapositions of their earlier work and the semi-smooth forms that the adoption of digital design brought into their work. It also reveals the deliberately awkward panelling that occurs when the form attempts to twist or fold, where the triangulations stick out like the bony scales on a lizard. It also demonstrates the functional adaptations that experimental digital work of the 1990s has undergone to become useful. In this construction, as well as in buildings by Gehry, Eisenman and others, the digital form-finding process runs in conjunction with simple physical modelling of programme (frequently consisting of foam models of windowless galleries and auditoria), and often results in a hierarchy of forms, from the simple internal spaces to the complex and iconic envelope, and with an evocative naming strategy accompanying the whole ensemble.

A new Mons International Congress Xperience, Belgium, enacts a number of similar urban gestures to the Confluences Museum: it performs a role as an urban focus in an area that has been recently redeveloped, connected closely to transport infrastructure, and it offers a public space that is beyond the immediate requirements of its brief.

The center, designed by Studio Libeskind with H2a Architecte & Associés, has been designed in a way that although it includes Libeskind's trademark formal gestures and motifs, it avoids the formal subservience that is sometimes associated with them. The building is organised around its primary programme of three auditoria of various sizes and a large open event hall, but it manages to include a large roof garden and a viewing platform for gazing back at the historic centre of Mons and various historic local landmarks.

Daniel Libeskind's approach to the building involved creating in-

蒙斯国际会展中心,比利时
Mons International Congress Xperience, Belgium

些线条在平面勾勒出杏仁的形状,从当地景观中跃然而出。这些结构的一侧是公园,另一侧是火车站。"带状"结构十分坚固,这是通过将其固定在具有两种相同色调(但质地不同)的钢条上来实现的:建筑下层为木条,上层则是香槟色铝板。为了使窗户不破坏"带状"形式的连续性,这些覆盖材料将跨过窗户,但会形成一个角度,不会阻挡人们的视野。

角度各异的钢框架,以及表面看起来相互连接的体量是李博斯金设计方式的全部。该建筑比他设计的其他大规模建筑要小一些,也简单一些,而其象征意义也相应的小一些,但该中心具有非常符合委托书要求的规模。人们对李博斯金的作品也产生过疑问。自20世纪90年代后期,李博斯金设计的柏林犹太人博物馆竣工后,李博斯金似乎与20世纪80年代解构主义理念的代表人物(如扎哈·哈迪德,蓝天组)背道而驰。他并不经常运用超越一切的复杂的建筑形式来建造建筑,而是依赖于结构并置、大角度体量,而非复杂的曲线所带来的幻象来设计。这就意味着这种建筑方法比那些具有象征意义的建筑方法更容易,且成本更低,而李博斯金的项目完全地反映了这些特点。但李博斯金是以设计一系列极度令人担心的、充满历史氛围的建筑物来开始他的职业生涯的。而人们看到这些建筑的视觉特色逐步地应用在各类建筑中,且横跨他的整个职业生涯,有时是很难理解的。

体现了文化建筑的新设计理念的还有位于西班牙塞雷纳新镇城郊的维加斯·阿尔塔斯会议中心和礼堂,由Pancorbo, de Villar, Chacón和Martín Robles设计。这个于近期竣工的项目展示了思考文化设计的一些新方向。乍一看,其形式主义仍然以蓝天组和丹尼尔·李博斯金的理念为主导,有着最简单的几何外形,且平面和立面围绕着效果来进行布局。但是该项目采用了一些不同的战略,使其看起来和一个价格昂贵的数码怪兽一样壮观。

法国里昂会议中心的大部分位于地下。景观横跨整个场地,位于屋顶,使建筑表面看起来比实际小很多,同时礼堂和其他大型空间复杂的几何外形也被隐藏起来。地平面向下插入土中,延伸至入口,屋顶的条形天窗模拟了附近田野的耕作形式。一座小塔成为该项目的主要建筑焦点,从平面来看呈方

terlocking "ribbons" that form almond shapes in plan, marked out by local landscape, with park on one side and railways to the other. The apparent solidity of the ribbons is created by fixing onto the steel frame strips of two materials of similar tone but very different qualities: timber slats for the lower levels of the building and a "champagne" colored aluminium for the upper levels. In order for the windows not to break the apparent continuity of the ribbon forms, the cladding material continues across them, although angled to allow views out.

The jarring angles of the steel frame and the apparently interconnecting volumes are all parts of the Libeskind design method, although the building is smaller and simpler than many of his more famous projects, and keeps its level of "iconicity" low. While this is appropriate to the scale of the commission, it does raise a question with regard to Libeskind's work. In the years since the Berlin Jewish Museum, completed in the late 1990s, Libeskind's work has followed a different path to some of the other members of the "deconstructivists" of the 1980s, such as Zaha Hadid and Coop Himmelb(l)au. A Libeskind building doesn't often involve exceedingly difficult forms to construct, instead usually relying on the illusion of juxtaposition, and extremely angular volumes rather than complex curves. This means that they are frequently more easy to construct than some other "iconic" architectural approaches, and thus more affordable, and Libeskind's commissions reflect this. But of course Libeskind began his building career with a series of extremely fraught historically charged projects, and seeing their visual characteristics gradually become applied to a wide variety of building types across his career has been difficult to understand at times.

A project which demonstrates some new directions in thinking about cultural projects is a recently completed Vegas Altas Congress Center and Auditorium on the outskirts Villanueva de la Serena, Spain, by Pancorbo, de Villar, Chacón and Martín Robles. At first glance, the formalism that dominates the thinking of the Coop Himmelb(l)au and Daniel Libeskind is largely reigned in, with the geometry mostly rather simple, organised around effects in plan and elevation. But the project utilises a number of different strategies to make for a building just as spectacular as an expensive digital monster.

Much of the center, far more than is the case for the building in

维加斯·阿尔塔斯会议中心和礼堂,西班牙
Vegas Altas Congress Center and Auditorium, Spain

形,有着混凝土外壳和圆角边缘,且设有各种不规则的洞口。塔内设有一个台塔,其功能是作为剧院以及咖啡厅、办公室和更衣室。

虽然其形式并不出挑,但其材料和细节却很引人注目。混凝土塔台的周围是轻质钢框架,一根很粗的编织绳利用上百条钢筋系于钢框架。建筑内很少使用绳子作为材料,而这种使用方式就更少见了;在这里,绳子看起来漂浮在建筑的周围,给人以整座建筑像一大捆干草的感觉。这种令人惊讶的细节处理方法也应用在建筑物内,混凝土天花板采用未经修饰的垂枝来装饰,这在20世纪60年代之后就很少见到;还有材质为聚碳酸酯和丙烯酸树脂的彩色背光装饰的墙体、玻璃增强石膏、曲形无框玻璃,它们共同形成了一种连续性,出位但材质细部处理地非常好。

一些西班牙的建筑师奉献了非常大胆的设计,他们巧妙地利用廉价的材料来展示他们的建筑,这些设计方法都非常符合21世纪的发展。除了Pancorbo事务所,还有一些事务所如Selgas Cano、Andres Jaque或Carlos Arroyo,它们都把典型的西班牙建筑中的施工技术与符合后网络时代和后简朴时代风格的材料使用方式和意象形成方法结合起来。这些事务所的建筑师在今后的几年内会逐步占据主导地位。

但是有时候一座优秀的公共项目既不采用特殊的、出位的形式,也没有使用明亮的材料。由NL建筑师事务所设计的Crossoverzaal音乐中心作为乌特勒支音乐宫的一部分,其外观几乎没有展现出来,但仍然是一个公共建筑佳作。

乌特勒支音乐宫是由过去的弗雷登伯格音乐中心改建而成的一组全新的音乐区。这座音乐中心是结构主义大师赫曼·赫茨伯格在20世纪70年代设计的作品。原建筑的大部分(除了主音乐厅)都被拆除,作为火车站周围重建区域的一部分,根据赫茨伯格的计划,场地内新建了一个大屋顶,四间新音乐厅位于其下,每间大厅都由不同的建筑师设计,一间是流行音乐厅,一间是爵士乐音乐厅,一间是室内音乐厅,最后一间是NL建筑师事务所设计的Crossoverzaal音乐厅,这是一处更加灵活的空间,兼具夜店、展厅和或其他用途。

Crossoverzaal音乐厅拥有整座建筑内最高的空间,而入口则是其最

Lyon, is under the ground. Across the site, landscaping has been applied to the roof level, allowing for the building to apparently appear far smaller than it might, but also to hide the difficult geometries of the auditoria and other awkward spaces. The ground level appears to fold down into the earth towards the entrance, with strips of rooflights mimicking the ploughing patterns of the nearby fields. The main architectural focus of the project is a small tower, square in plan, with a concrete shell, rounded at the edges and perforated with a variety of irregular openings. This contains the fly tower for the theatre space as well as a cafe, offices and changing rooms.

While the form might not be particularly bold, the materials and detailing certainly are. Surrounding the concrete fly-tower building is a very lightweight steel frame, and tied to that by hundreds of reinforcing bars is a thick woven rope. This rope, a rare enough material in architecture, let alone used in this way, appears to float around the building, appearing like "a giant bale of hay". This startling approach to detail is continued within the building, where the concrete ceilings have a rough "weeping" finish not often seen since the 1960s. Elsewhere, colourful polycarbonates, acrylics, back-lit walls, glass reinforced gypsum and curved frameless glazing make up a riot of loud but well detailed material connections.

This approach is very much a 21st century development, with a number of Spanish architects offering extremely bold, clever uses of often cheap materials to make their architectural statements. Alongside Pancorbo there are firms such as Selgas Cano, Andres Jaque, or Carlos Arroyo, all of whom combine the construction skills typical of a Spanish architectural training with a more post-internet, post-austerity approach to materials and imagery. These architects are likely to become more and more prominent over the coming years.

But sometimes it is neither a specific formal boldness nor a bright use of material that makes for a prominent public project, and the Crossoverzaal, a concert space designed by NL Architects as part of the Utrecht Music Palace, has almost no external presence, yet represents a remarkable commission for a public building.

The Utrecht Music Palace is a new set of concert venues that have been built out of the old Vredenburg music center, a 1970s building by structuralist Herman Hertzberger. Much of the original

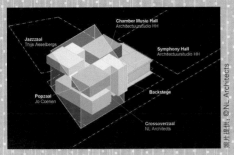

乌特勒支音乐宫的空间布局
space organization of Utrecht Music Palace

低点,地面高0.5m。每间大厅都是隔音的,其基本的规划是盒子中的盒子。NL建筑师事务所决定扩建空间,并在其内设置流线,同时使大厅本身进行了延伸。最矮的空间是门厅,人们站在大型落地窗前可俯瞰城市,另外还有一组楼梯可直接通向大厅。拾级而上,人们还有可能来到几处阳台区。

细节设计最小化,内部和外部都尽量空旷,礼堂如同一个黑盒子,门厅和外部表面为白盒子。但是正是大厅的这种独特的氛围造就了如此出位的前景。整个大厅悬在空中,其额外的立面,即底面,连同各个突出的结构以及悬臂结构使建筑极具雕塑感。

在整个乌特勒支音乐宫里,Crossoverzaal音乐厅显得更加突出。该建筑综合体在建筑方面并不显得精致,但是其商业功能却进行了精心的设计,且其功能理念尤为具有挑战性。通过在内部建造一处内设公共区域和不同建筑(由不同的建筑师设计)的大型空间,一个20世纪60年代才有的超大结构便展现在人们面前,这是一个巨大的娱乐大厅,建筑内部人声鼎沸,喧嚣吵杂。

很早就有人预言了它的衰落,特别是在近十年的金融危机中,但是这个巨型的标志性建筑时至今日仍是人来人往,颇受欢迎。建筑师自有办法使公共建筑彰显大胆、明亮以及充满活力的特色,也使下一代的建筑能够少在外形、技术方面进行试验,而是更多地研究材料、色彩和结构并置方面。

building (apart from the main concert hall) was demolished as part of a redevelopment of an area around the train station, and in its place, according to a plan by Hertzberger, would be a giant roof under which there would be four new concert halls. Each of these halls would be handled by a different architect, with one dedicated to popular music, one for jazz, one for chamber concerts, and the Crossoverzaal, a more flexible space capable of being used as a nightclub, an exhibition hall or other uses, designed by NL Architects.

The Crossoverzaal is one of the highest spaces in the building, with its lowest point, the entrance, already at level 05. Each of the halls has to be acoustically isolated, so the basic programme was a box within a box. NL Architects decided to enlarge this space and inhabit it with both circulation and extensions to the the hall itself. The lowest space is a foyer, with a picture window looking out over the city rooftops, and then another set of stairs lead up to the hall. From there it is possible to rise further up to a number of balcony areas.

Detailing is minimal - the interior and exterior are as blank as possible, the auditorium as a dark box, and the foyer and outside surface as white boxes. But it is the unique context of the hall that allows it to be so such a bold prospect. By being hoisted up into the air, the hall has an extra facade - its underside - and the different projections and cantilevers mean that it gives off a highly sculptural effect. In the context of the Utrecht Music Palace project as a whole it is even more remarkable. The complex may not be architecturally refined, in fact it's rather on the commercial side, but the functional ideas within are still challenging. By creating a huge indoor space with a public area and different buildings within, designed by different architects, there is something of the 1960s era of megastructures about the project, a vast hall of entertainment, with a cacophony of different architectures for the different activities inside. Its decline has been predicted for a long time, especially in a world after the financial crash of late last decade, but the popularity of the giant icon building with clients continues to this day. But there are definitely other ways that an architect can be bold, bright, and dynamic in their approach to public architecture, and it's possible that the next generation of these buildings will be less experimental in terms of shape and technique, and more so in materials, colors and juxtapositions. Douglas Murphy

汇流博物馆
Coop Himmelb(l)au

另辟蹊径的文化建筑 Alternative Strategies for Culture Buildings

自2001年法国里昂举办了关于自然历史博物馆的国际建筑竞赛以来,博物馆就肩负了"传递知识的媒介"的重任,而不再是商品的陈列室。

该博物馆的场地位于罗讷河和索恩河汇流处的一座半岛上,这里在100年前曾被人们扩建。该项目看上去是一个艰难的任务(536m²长的桩必须牢固地嵌入地下),但是这个场所对于城市设计来说显然是至关重要的。对于从南侧进入的参观者来说,建筑是一座与众不同的灯塔和入口,同时也是城市开发的起点。

为了建造一座传播知识的博物馆,这座拥有新形式的综合设施必须设有一个标志性的门面。建筑必须依靠新颖的几何外观才能脱颖而出,设计理念中有一点非常重要,就是从城市来此的客流不应该被一座建筑阻挡。因此,建筑师带着这样的理念,设计了一座开放式的、可穿透的建筑,只依靠一些柱子的支撑使部分浮在空中,以在其底部创造一处公共空间。

建筑基本包括三个部分,包括位于微微升起的地基之上(由于高水位的原因)的两座礼堂(可分别容纳327人和122人),以及一些工作区,这些工作区同时也用作周边学校的培训区。这三个部分都会建在生产展品的仓库和工作坊旁边。

入口建筑被称为"水晶体",是开放式的、可穿透的结构,也是通往展览区的一个垂直入口。此外,人们经由一座电梯、一座楼梯和一条螺旋形坡道还可以到达被称为Espaceliant的连接步道。步道左右两侧分布着一些私人展览厅(有一个是两层结构),而步道的尽头是两条河流的交汇处,即汇流点。钢结构呈现为桥梁的形态,实现了所有展厅都处于无支撑状态的设想。行政办公室位于展览空间的上层。

在急速上升区域下方的广场内,一些元素呈飞翔的姿态(部分展厅为悬挑形式)。光波在小型湖面上形成美丽的图案,然后又反射到建筑的底部。一间啤酒吧强调了该区域的公共性。此外,参观者还可以轻松到达位于顶层的露台咖啡室。

在建筑入口处,水滴形结构起到了支撑的作用。其建筑形式的设计灵感来源于两汨水流汇聚形成的湍流。自流井有效减少了入口建筑整体钢架结构三分之一的重量。

Confluences Museum

Right from the 2001 international competition for a natural history museum in Lyon, the museum was envisioned as a "medium for the transfer of knowledge" and not as a showroom for products. The building ground of the museum is located on a peninsula that was artificially extended 100 years ago and situated in the confluence of the Rhône and Saône rivers. Even though it was apparent that this site would be a difficult one(536m piles had to be securely driven into the ground), it was clear that this location would be very important for the urban design. The building should serve as a distinctive beacon and entrance for the visitors approaching from the south, as well as a starting point for urban development. In order to build a museum of knowledge, a complex of new form had to be developed as an iconic gateway. A building that truly

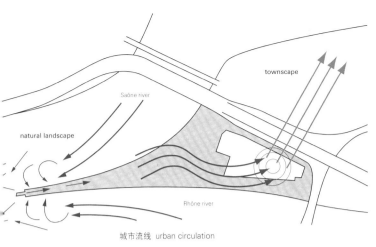

城市流线 urban circulation

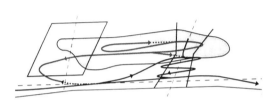

水晶外形 crystal

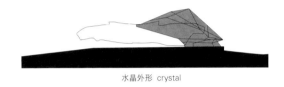

云外形 cloud

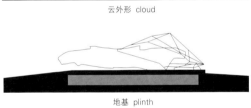

地基 plinth

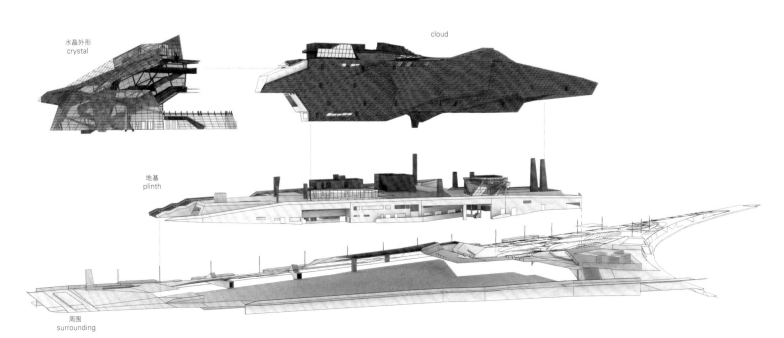

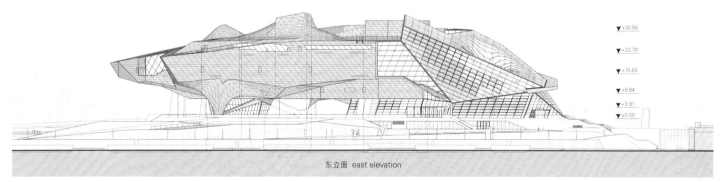

东立面 east elevation

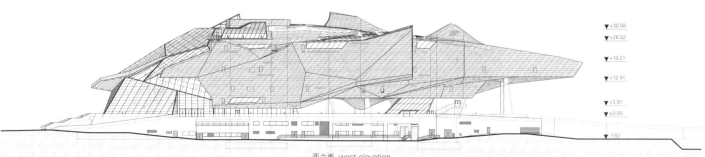

西立面 west elevation

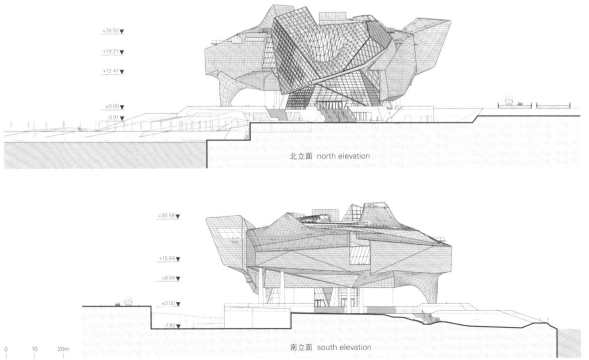

北立面 north elevation

南立面 south elevation

stands out can only come into being through shapes resulting from new geometries. It was important to the concept that the flow of visitors arriving from the city to the Pointe du Confluent should not be impeded by a building. The idea was therefore to develop an openly traversable building that would be floating in part only on supports, in order to create a public space underneath.

Essentially, the building consists of three parts. Situated on a slightly raised base(due to the high groundwater), two auditoriums(for 327 and 122 persons, respectively) and work spaces, which will also be used for training purposes for the surrounding schools, will be located next to storage and workshops for the production of exhibitions.

The entrance building, the so-called Crystal, is openly traversable, and is a vertical access to the exhibition spaces. The so-called Espace liant, a connecting path, can be reached by an escalator, a staircase, and a spiral ramp. Left and right of this path are arranged by the individual exhibition halls(one of them is two-level), and at the end is a view of the confluence of two rivers, the Pointe du Confluent. The steel structure, conceived as a bridge construction, made it possible to develop all of the exhibition halls without supports. The administration rooms are located above the exhibition spaces.

In the Plaza below this highly raised, almost flying component – the showrooms are broadly cantilevered in parts – the lit wave pattern of the surface of a small lake is reflected on the underside of the building. A brasserie emphasizes the public nature of this place. A freely accessible terrace cafe is located on the top floor.

In the entrance building, a drop-shaped construction serves as the supporting structure. Its form was developed out of the turbulent flow created by the confluence of the two streams. This gravity well reduces the weight of the entire steel structure of the entrance building by a third.

A-A' 剖面图 section A-A'

1. 水池 — 1. water basin
2. 礼堂 — 2. auditorium
3. 大堂 — 3. hall
4. 工作室 — 4. atelier
5. 办公室 — 5. office
6. 图书馆 — 6. library
7. 自助厨房 — 7. self service kitchen
8. 水晶形结构流线区 — 8. crystal circulation area
9. 接待处 — 9. reception
10. 地铁入口 — 10. sub entrance

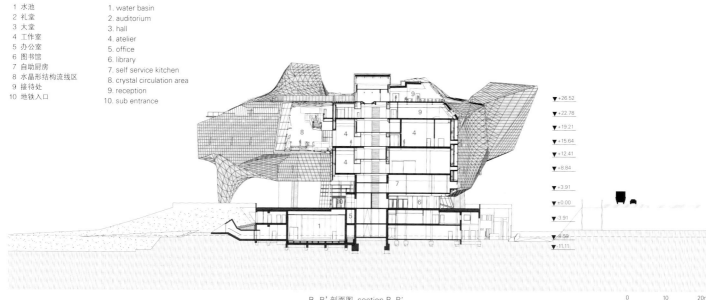

B-B' 剖面图 section B-B'

水晶形结构规划开发模型
structural scheme development modeling of crystal

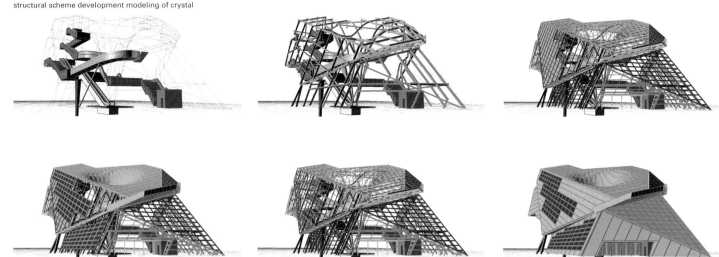

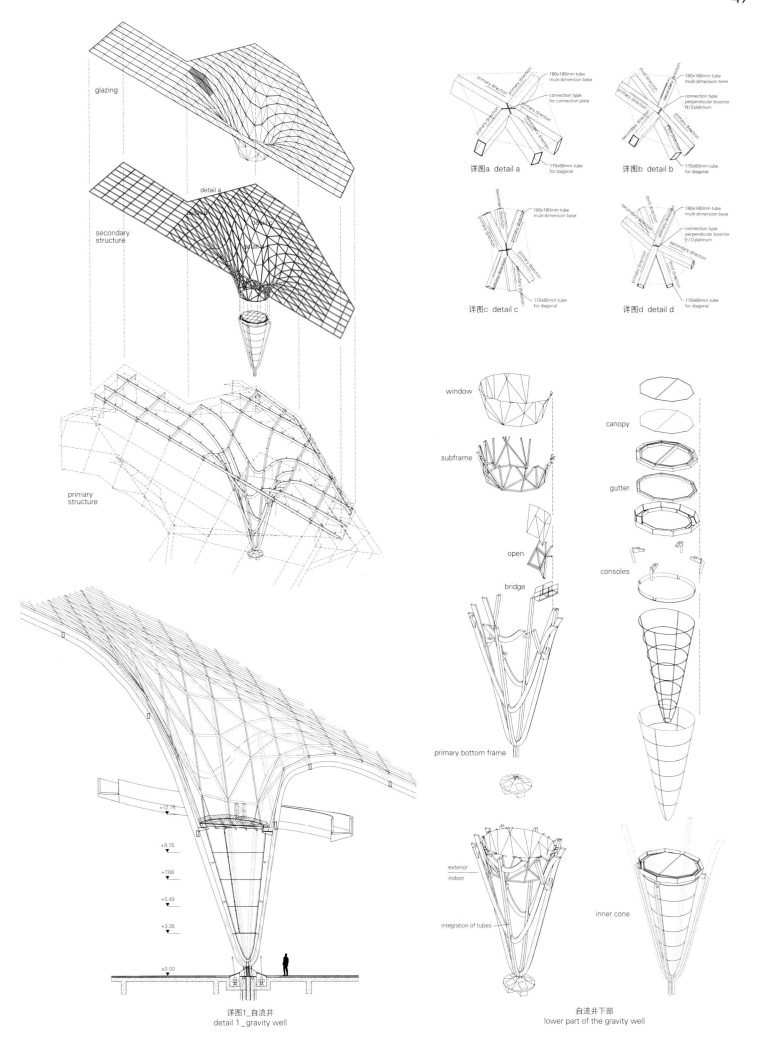

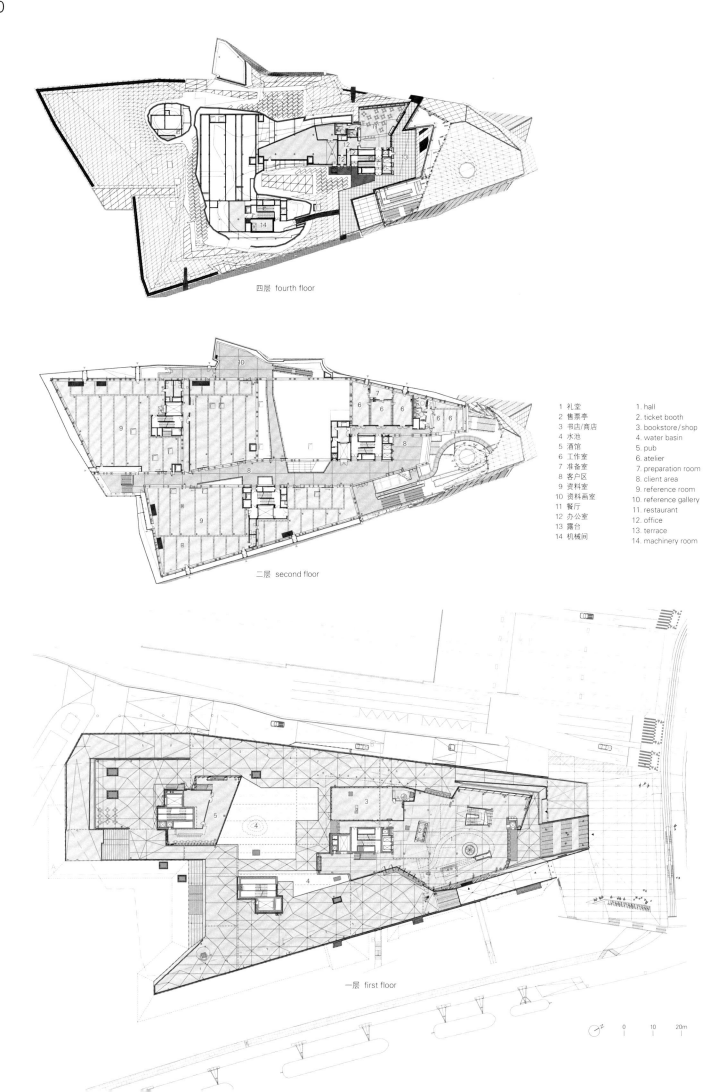

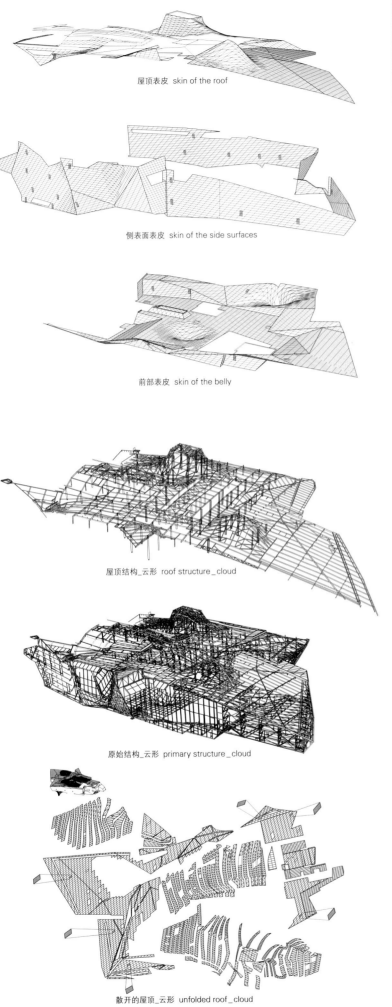

屋顶表皮 skin of the roof

侧表面表皮 skin of the side surfaces

前部表皮 skin of the belly

屋顶结构_云形 roof structure_cloud

原始结构_云形 primary structure_cloud

散开的屋顶_云形 unfolded roof_cloud

项目名称：Confluences Museum / 地点：86 quai Perrache, 69002 Lyon, France
建筑师：Coop Himmelb(l)au
设计总监&CEO：Wolf D. Prix / 项目合作者：Markus Prossnigg
建筑设计师：Tom Wiscombe / 项目建筑师：Mona Bayr, Angus Schoenberger / 项目协调师：Thomas Margaretha, Peter Grell
本地建筑师：planning_ Patriarche & Co, execution_ Tabula Rasa, project management_ Chabanne & Partenaires
项目团队：Vienna_ Christopher Beccone, Guy Bébié, Lorenz Bürgi, Wolfgang Fiel, Kai Hellat, Robert Haranza, Alex Jackson, Georg Kolmayr, Daniel Kerbler, Lucas Kulnig, Andreas Mieling, Marianna Milioni, Daniel Moral, Jutta Schädler, Andrea Schöning, Mario Schwary, Markus Schwarz, Oliver Tessmann, Dionicio Valdez, Philipp Vogt, Markus Wings, Christoph Ziegler, Lyon_ Patrick Lhomme, Francois Texier, Philippe Folliasson, Etienne Champenois, Alexandru Gheorghe, Niels Hiller, Emanuele Iacono, Pierre-Yves Six
施工测量：Jean Pierre Debray / 造价顾问：Mazet & Associés / 结构工程：design_ B+G Ingenieure, Bollinger und Grohmann GmbH, execution_ Coyne et Bellier, VS_A
HVAC：ITEE-Fluides / 消防顾问：Cabinet Casso & Cie / 音效顾问：Cabinet Lamoureux / 媒体顾问：Cabinet Labeyrie
照明顾问：Har Hollands / 景观设计：EGIS aménagement
甲方：Département du Rhône
用地面积：20,975m² / 有效楼层面积：46,476m² / 占地面积：9,300m²
造价：EUR 3,980/m²
竞赛时间：2001 / 施工时间：2006 / 竣工时间：2014.12
摄影师：
©Sergio Pirrone (courtesy of the architect)-p.40~41, p.43 middle, p.48, p.51, p.52, p.53, p.54, p.55
©Duccio Malagamba (courtesy of the architect)-p.43 top, bottom, p.44~45, p.47, p.56~57

另辟蹊径的文化建筑 Alternative Strategies for Culture Buildings

蒙斯国际会展中心
Studio Libeskind + H2a Architecte & Associés

位于比利时蒙斯市的蒙斯国际会展中心由李博斯金建筑工作室与当地的H2a建筑公司联合设计，CIT Blaton/Galère为总承包商。该中心于2015年1月9日面向公众开放，同时也拉开了"2015·欧洲文化之都"的序幕。

新会展中心被设想为蒙斯市的一座新建筑地标，是该地区经济振兴计划的关键一步，并成为新旧之间的衔接点。参观者能够从楼顶的观景平台远眺到位于城镇的历史中心位置、被联合国教科文组织列为遗址的17世纪的钟楼塔、圣地亚哥·卡拉特拉瓦设计的新火车站以及La Haine河。

面积为12 500m²的会展中心的几何表现形式具有鲜明的对比性。该中心设有一个宏伟的入口大厅、三座礼堂、一处多功能活动厅、会议室、办公室、餐厅、地下停车场以及公共屋顶露台。

"这栋建筑如同一块温润的宝石，我们运用简单却引人注目的设计形态、当地的材料和灵活的功能程序来装点它。"丹尼尔·李博斯金如此阐释，并补充道，"我们希望这个新的会展中心能为蒙斯市的振兴带来新鲜的活力"。

会展中心的覆层形成了外部纹理，并且为室内结构提供照明。倾斜的香槟色的电镀铝质带状外墙将建筑包裹起来，顶端从北侧的街道悬挑出来。而外墙下部覆以未经修饰的洋槐木制成的垂直板条，以呼应一旁的公园里的树木。从平面图来看，会展中心整体呈螺旋状升起，并且连接覆有绿色植被的屋顶和公共露台。

北侧的立面与街道处在同一平面，呈略微上升态势，以形成一个玻璃的入口，饰以深蓝色的铝框。

为了保持建筑形式方面的视觉整体性，建筑师沿着建筑带状的墙壁嵌入了一些窗户。这些窗户前方为垂直板条，而板条环绕着建筑，使照明和观景性能良好。综合设施的四周是抛光的浅灰色混凝土铺成的前院，内嵌的比利时蓝色岩石带产生斑驳的效果。这些蓝色的岩石带一直延伸至带状的墙壁上，并延伸至室内，形成不规则的图案，以保持设计的一致性。

进入综合体内部，迎接参观者的将是一处明亮且开放的双高大堂或者"论坛"。日光透过天窗照射进来，将新月形的空间切割成不规则的区间，创造出灵动狡黠的图案。暖灰色的混凝土地面镶嵌了纵横交错的比利时蓝色岩石带，充满雕塑感的大楼梯使用了现浇混凝土，饰以亮白光洁的表面，而镶以比利时蓝色岩石扶手的阶梯将引领参观者到达上层的礼堂。

该中心包括三座不同规模的礼堂，分别容纳500、200和100个座席，每座礼堂均安装了丹尼尔·李博斯金为Poltrona Frau/Cassina设计的充满活力的橙色七巧板坐椅。除了能举办一些特殊活动、派对和临时展览的论坛，中心还有380m²的多功能空间，以及16间大小不一、灵活布局的会议室。

建筑还具有能源高效性，配以绿化屋顶、被动遮阳系统、夜间制冷系统以及太阳能电池，因此获得Valideo B级认证，这相当于比利时的LEED金奖认证。"对我来说可持续建筑不是一种新的设计趋势或一种额外的特色，它是一种常识，也是一种品质。如果你运用了正确的建筑方法（通过使用高品质的材料、创新的方法与科技），你自然能够创造出持久、可持续的建筑"，丹尼尔·李博斯金说。

Mons International Congress Xperience

Mons, Belgium – designed by Studio Libeskind in collaboration with local partner H2a and general contractor CIT Blaton/Galère, the Mons International Congress Xperience opens to the public and kicks-off the Cultural Capital of Europe 2015, on January 9.

Conceived as a new architectural landmark for Mons, the new convention center is a key element in a plan for economic revitalization, and serves as a connector between the old and the new. From the viewing platform at the top, a visitor can spy the 17th-century Belfry tower, a UNESCO Heritage Site, in the historic center of town, a new train station designed by Santiago Calatrava and the La Haine River.

The 12,500m² center is an expression of contrasting geometric forms. The center houses a grand entrance hall, three auditoriums, a multi-purpose event hall, conference rooms, offices, a restaurant, an underground parking and a public roof terrace.

"We used simple, yet dramatic, design gestures, local materials and a flexible program for this modest gem of a building", said Daniel Libeskind." We hope the new center brings a fresh dynamic to this area of revitalization in Mons", adds Libeskind.

Clad in a manner that gives texture and light to the structure,

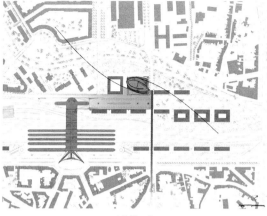
中轴线 axis

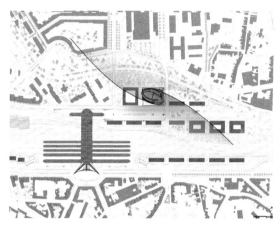
螺旋形布局 spiral layout

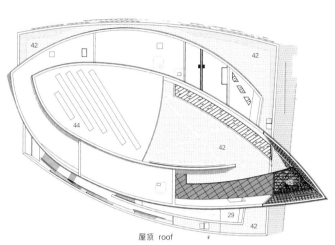
屋顶 roof

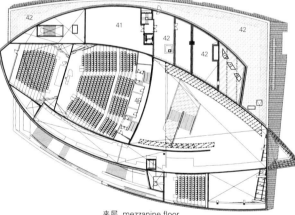
夹层 mezzanine floor

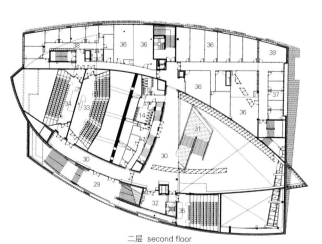
二层 second floor

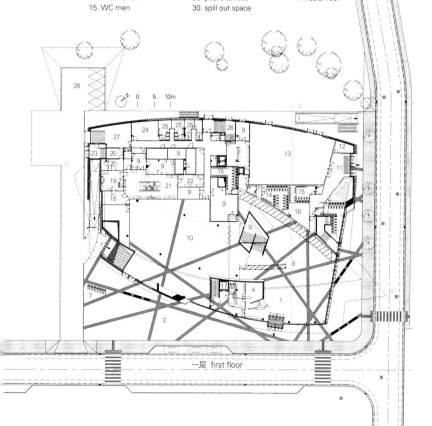
一层 first floor

1 入口大厅
2 广场
3 前台
4 代客存衣处
5 安保处
6 VIP大堂
7 自行车存放处
8 论坛/展览空间
9 存储室
10 餐厅/可伸缩空间
11 大堂
12 接待处
13 活动空间
14 女士卫生间
15 男士卫生间
16 零售店
17 通往公共露台的通道
18 退餐区
19 餐饮区
20 废品区
21 厨房/预备室
22 工作台
23 员工入口
24 工作室
25 更衣室
26 自行车存储室
27 装卸平台
28 员工停车场
29 公共露台
30 疏散区
31 大楼梯
32 美发沙龙
33 礼堂-1(500座)
34 礼堂-2(200座)
35 礼堂-3(100座)
36 会议室
37 办公室
38 行政区
39 休息室
40 控制室
41 车间
42 绿色屋顶
43 瞭望塔
44 太阳能屋顶

1. entrance hall
2. plaza
3. front desk
4. coat check
5. security
6. vip lobby
7. bike stands
8. forum/exhibition space
9. storage
10. restaurant/flex room
11. lobby
12. reception
13. event space
14. WC women
15. WC men
16. retail
17. access to public terrace
18. dishes return
19. dishes
20. waste
21. kitchen/prep room
22. staging
23. staff entrance
24. workshop
25. changing rooms
26. bike storage
27. loading dock
28. staff parking
29. public terrace
30. spill out space
31. grand stair
32. salon
33. auditorium-1(500seats)
34. auditorium-2(200seats)
35. auditorium-3(100seats)
36. meeting rooms
37. offices
38. administration
39. break room
40. control rooms
41. plant room
42. green roof
43. belvedere
44. solar roof

东南立面 south-east elevation

东北立面 north-east elevation

西北立面 north-west elevation

西南立面 south-west elevation

canted ribbon walls of champagne, anodized aluminum wrap the form upwards to a prow that cantilevers over the street to the north. The lower walls are clad with vertical slats of unfinished Robinia wood that echo the trees in a neighboring park. In plan, the center is a spiral that ascents on its self and articulates a planted green-roof and public terrace.

To the north, at street level the facade appears to lift up to reveal a glazed entrance, finished with deep blue aluminum mullions. To maintain the visual integrity of the form, the architect inserted few windows along the ribbon wall – and those are fronted by slats that are rotated to allow for daylight and views. Surrounding the complex is a forecourt of polished, light grey concrete, flecked with bands of Belgium blue stone. These blue bands continue onto the ribbon walls and into the interior forming an irregular pattern that unifies the design.

Visitors enter the space to encounter a bright, open double-height lobby or "Forum". Here skylights cut through the length of the crescent-shaped space at irregular intervals to usher in daylight, and create shifting patterns of natural illumination. Soft grey concrete floors are cross-hatched with inlaid Belgium blue stone, a sculptural grand staircase is constructed of cast-in-place concrete and finished with a white gloss surface, and blue stone steps lead visitors to the upper auditorium floors.

The Center features 3 auditoriums of varying sizes: 500; 200; and 100 seats, each fitted with vibrant orange Tangram seats that Daniel Libeskind designed for Poltrona Frau/Cassina. In addition to the Forum, which can host special events, parties and temporary exhibits, the Center features a 380m² dedicated multi-event space as well as 16 meeting rooms of varying sizes with flexible layouts.

Highly energy efficient, with a green roof, passive shading, night cooling and fitted with photovoltaic cells, the MICX is on track to achieve Valideo Status B, Belgium's equivalent to LEED gold. "For me a sustainable building is not a new design trend or an added feature, it is about common sense and quality. If you build in the right way using high-quality materials, innovation and technology, you will create something long lasting and sustainable", said Daniel Libeskind.

项目名称：Mons International Congress Xperience
地点：Avenue des Bassins, 7000 Mons, Belgium
建筑师：Studio Libeskind, H2a Architecte & Associés
结构工程师：Ney & Partners
电气工程师：Putman
机械工程师：CIT Blaton
总建筑面积：12,500m²
造价：EUR 27,000,000
竞赛时间：2010
施工时间：2012.6
竣工时间：2015.1
摄影师：©Georges De Kinder (courtesy of the architect) - p.66, 68, 69[top]
©Hufton+Crow (courtesy of the architect) - p.58~59, 60~61, 63, 64~65, 67, 69[bottom]

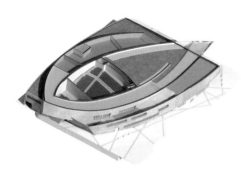
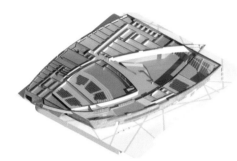

1 Melina Mercouri大道 2 广场 3 论坛 4 零售店 5 活动空间 6 公园 7 停车场 8 会议室 9 机械区 10 绿色屋顶 11 瞭望塔 12 公共露台
1. Melina Mercouri Avenue 2. plaza 3. forum 4. retail 5. event space 6. park 7. parking 8. meeting rooms 9. mechanical area 10. green roof 11. belvedere 12. public terrace
A-A' 剖面图 section A-A'

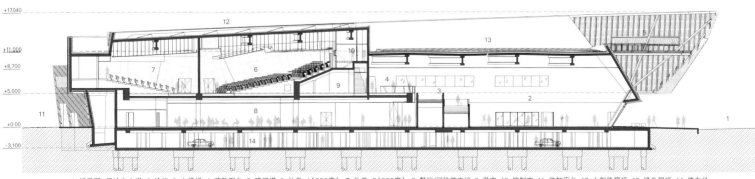

1 托马斯·爱迪生大道 2 论坛 3 大楼梯 4 疏散阳台 5 瞭望塔 6 礼堂-1(200席) 7 礼堂-2(500席) 8 餐厅/可伸缩空间 9 温室 10 控制室 11 装卸平台 12 太阳能屋顶 13 绿色屋顶 14 停车场
1. Thomas Edison Avenue 2. forum 3. grand stair 4. spill out balcony 5. belvedere 6. auditorium-1(200seats) 7. auditorium-2(500seats) 8. restaurant/flex room 9. green room 10. control room 11. loading dock 12. solar roof 13. green roof 14. parking
B-B' 剖面图 section B-B'

乌特勒支音乐宫

architectuurstudio HH + NL Architects + Jo Coenen Architects & Urbanists + Thijs Asselbergs

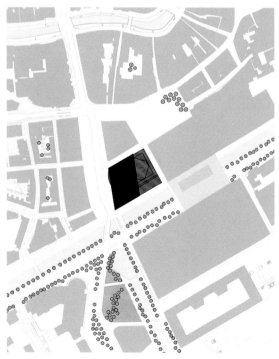

乌特勒支城将1978年建造的、带有主礼堂的原音乐中心改造成为一座音乐宫，额外内设一个音乐厅，用于演奏流行音乐、室内音乐、爵士乐以及跨界音乐作品。弗里登堡音乐中心是Hoog Catharijne大型购物中心的一部分，但这座新的音乐宫完全是一座独立的音乐建筑。

该项目的设计理念是在原有的弗里登堡音乐中心之上建造一座音乐建筑，内设四座大厅。这意味着将不得不拆除Hertzberger的部分杰作。该音乐宫将不同的目标群体聚于一座建筑之中：除交响音乐厅之外，这座综合体内的一座可容纳2000人的大厅专门用于演奏流行音乐，此外，它还设有一间爵士乐厅，一间室内音乐厅和一间被称为Crossoverzaal的跨界音乐厅。这些"音乐人的聚居地"在architectuurstudio HH事务所制定的总体规划下分别由四位建筑师完成。

Architectuurstudio HH事务所意在打造一张大约50m长的"大桌子"，横跨整个礼堂区，而新建的音乐厅及其门厅叠加于其上。

整体结构被正方形的穿孔覆层包裹起来，覆层在夜晚如同一个巨大的立方体灯饰闪耀在夜空，灯光同时也照亮了建筑内部。25m高的"桌面"组成了50m×50m的城市广场，人们在此能够欣赏到宏伟壮阔的城市景观。此外，这里的音乐厅和休息厅还设有咖啡馆，甚至有可能设置一个餐厅。

NL建筑事务所设计的Crossoverzaal厅作为承办广泛类型演出的独立大厅,可用于舞会、脱口秀、时尚秀、产品展览和会议。但它还有一项重要的功能是作为安全阀,在其他厅内发生无法应对的特定事件时发挥作用。该厅包括一间大型休息厅和一间地面十分平坦的厅供灵活使用。这里没有固定的舞台或座位,基础配置只有400个坐席,但最多可容纳600人。更衣室和后台区离得很近,便于重新布局。

音乐厅本身如同一个"盒子"。阳台和"凉廊"通过不寻常的方式与之连接,以突出主体量。音乐厅开创了一条大胆的路线,而门厅悬于其下。

设计出发点是"盒中盒"原则。建筑内外层之间的声腔被扩大了,以满足建筑、基础设施及其附加功能的需求。

建筑的中心理念是联系。Crossoverzaal厅连接其他几个厅,向四面八方延伸,形成星形结构。主入口在流行音乐厅的顶部。其中一个阳台直接通到爵士音乐厅。跨界音乐厅与室内音乐厅的门厅都与同一个区域连通。一个"触手"式结构与屋顶连通,将日光引入厅内,而另外两个同样的结构内设酒吧和阳台,能够接触到外立面,面向城市上空的壮观景象展开怀抱。

Utrecht Music Palace

The city of Utrecht intends to transform the 1978 Music Center with its existing main auditorium into a Music Palace with an additional hall for pop, chamber music and jazz as well as cross-over music. The Music Center Vredenburg was part of the large Hoog Catharijne shopping center, but the new Music Palace is free-standing building dedicated entirely to music.

The idea was to construct a music building with 4 halls on top of the existing concert hall Vredenburg. That would mean to partly demolish the Hertzberger Masterpiece. It also would bring together totally different target groups in one building: in addition to the symphony hall, the complex will feature a 2,000 people hall dedicated to pop music, a jazz hall, a room for chamber music and the so-called Crossoverzaal. These "Biotopes" will be developed by four different architects within the master plan by architectuurstudio HH.

Architectuurstudio HH proposes to construct a "table" of approximately 50 meters, spanning the existing auditorium block, on which the new additional halls with their foyers can be stacked. The entire structure will then be encased within a square perforated cover which will shine out at night like a huge cubiform lamp, as well as let light shine in the interior. Some 25 meters high, the "table top" forms a 50 x 50m city square with superb city views. Besides concert halls and foyers it will also house cafes and possibly a restaurant.

The Crossoverzaal of NL Architects will serve as an independent hall that can be used for a wide range of performances. It can be used for dance parties, stand-up comedy, fashion shows, product presentations and congresses. But it can also function as a "safety valve", when other rooms lack capacity for certain events. This biotope includes a spacious foyer and a flexible hall with a flat floor. It doesn't have a fixed stage or fixed seating. The basic configuration holds 400 seats, but it can contain up to a maximum of 600 people. The proximity of dressing rooms and back stage areas allows for numerous arrangements.

The Hall itself is a Box. Balconies and "loggias" are connected to it in an irregular way, bulging out of the main volume. The biotope creates an adventurous routing, and the foyer is suspended under it.

The starting point is the box-in-box principle. The acoustic cavity in-between the inner and outer layer is enlarged to fit the building's infrastructure and additional programs.

The central idea is to connect. The Crossoverzaal is reaching out to the other halls, stretching out in all directions, creating a kind of star shape. The main entrance is on top of the Pop. One of the balconies creates a direct link to the Jazz Hall. The foyer of Crossover and Chamber music hall can be connected into one space. One tentacle connects to the roof to allow daylight in the hall. Two others, containing a bar and a balcony, touch the facade and open up to spectacular views over the city.

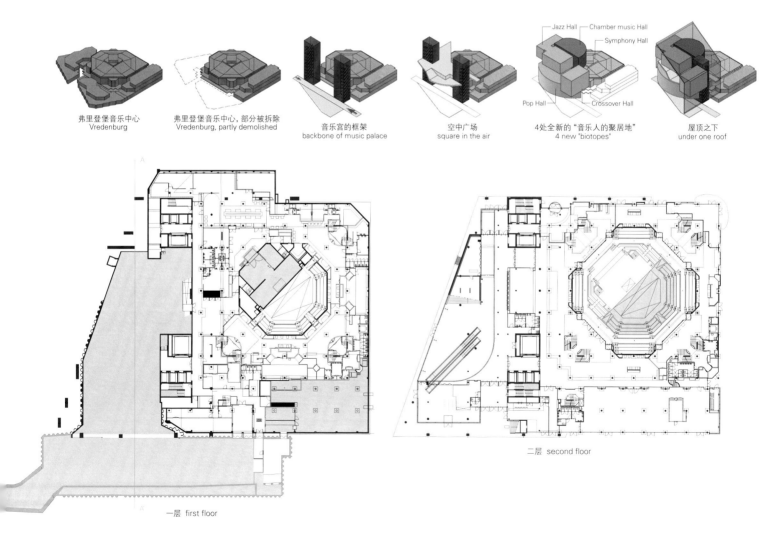

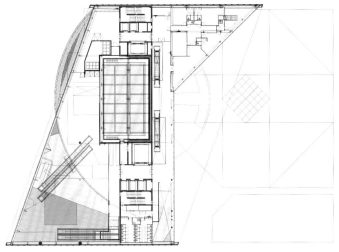

七层 seventh floor

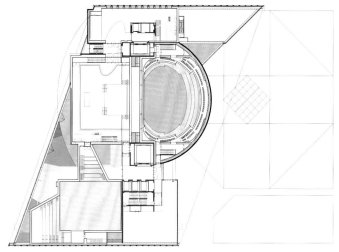

十一层 eleventh floor

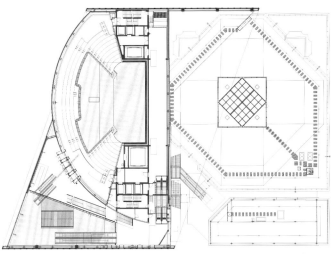

六层 sixth floor

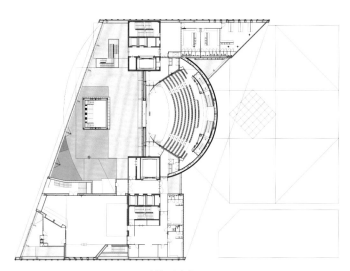

九层 ninth floor

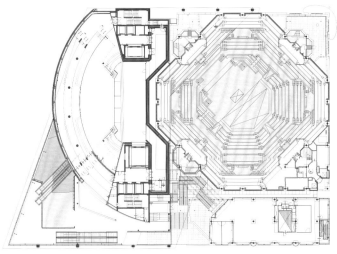

四层 fourth floor

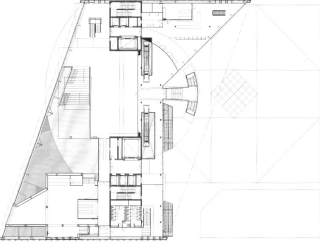

八层 eighth floor

项目名称：Music Palace Utrecht
地点：Tivoli, Vredenburg Utrecht, Netherlands
建筑师：
Master plan and Coordination _ architectuurstudio HH
Symphony Hall, Chamber Music Hall _ architectuurstudio HH
Crossoverzaal _ NL Architects
Popzaal _ Jo Coenen Architects & Urbanists
Jazz Hall _ Thijs Asselbergs
甲方：Muziekcentrum Vredenburg
造价：EUR 75,000,000
面积：31,000m²
设计时间：2003
施工时间：2009
竣工时间：2014
摄影师：©Luuk Kramer (courtesy of the architect)

平面概念图
plan concept

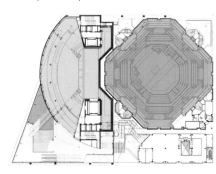

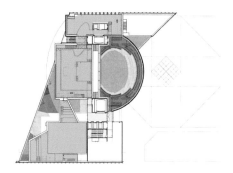

剖面概念图
section concept

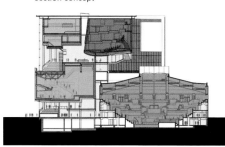

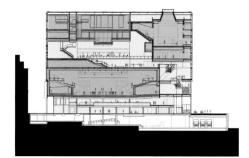

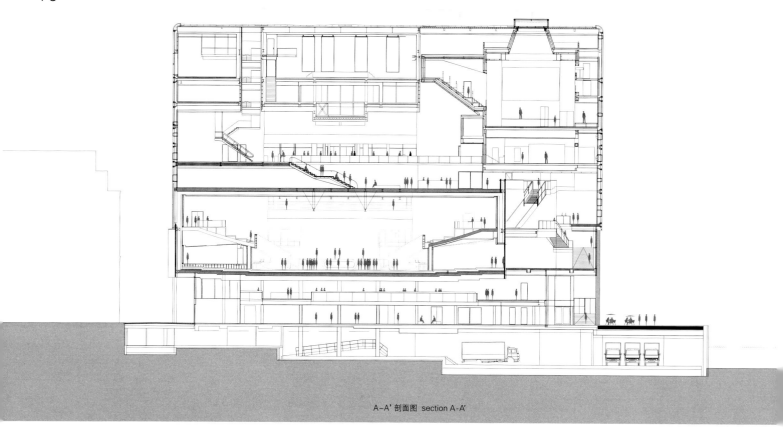

A-A'剖面图 section A-A'

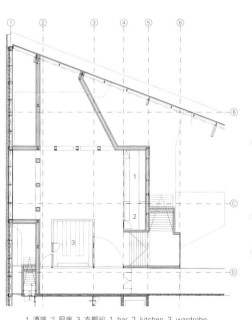
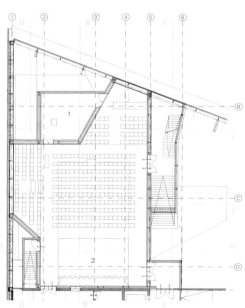
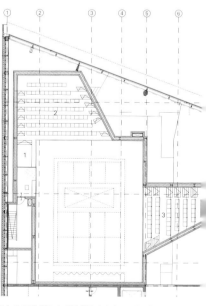

1 酒吧 2 厨房 3 衣帽间 1. bar 2. kitchen 3. wardrobe
七层上层_Crossoverzaal音乐厅
seventh upper floor _ Crossoverzaal Hall

1 储藏室 2 舞台 1. storage 2. stage
八层_Crossoverzaal音乐厅
eighth floor _ Crossoverzaal Hall

1 阳台 2 78坐席区 3 47坐席区 1. balcony 2. 78seats 3. 47sea
十层_Crossoverzaal音乐厅
tenth floor _ Crossoverzaal Hall

B-B' 剖面图 section B-B'

C-C' 剖面图 section C-C'

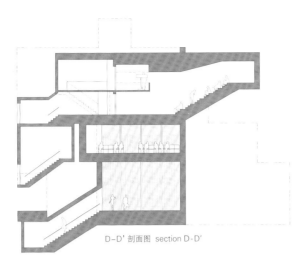
D-D'剖面图 section D-D'

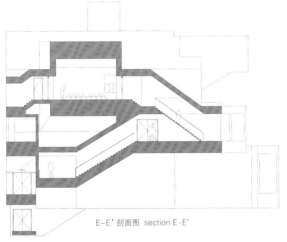
E-E'剖面图 section E-E'

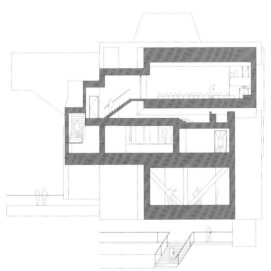
F-F'剖面图 section F-F'

G-G'剖面图 section G-G'

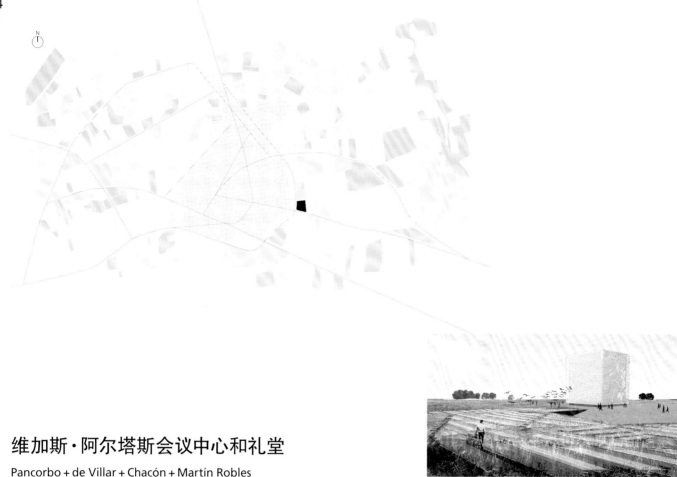

维加斯·阿尔塔斯会议中心和礼堂

Pancorbo + de Villar + Chacón + Martín Robles

维加斯·阿尔塔斯会议中心和礼堂位于城乡边缘的模糊地带。建筑旨在使维加斯区内这座独立的建筑的永恒环境脱颖而出。它像一捆巨大的干草浮于乡间，周围是平坦的地平线，土地漫无边际且富饶。这个嵌入结构给予了小镇广阔的绿色公共空间，一个大型绿色屋顶覆盖了大部分的功能空间和展厅。该立方体建筑包裹在绳子编织的网内，是周围唯一可以看到的地面建筑，既满足了所需的可见度，又将为舞台而建的大型体量隐藏起来。建筑真正的立面由条状的植物构成，这些植物为粗细度一致的绿色稻草，而建筑表面则被照亮地下空间的天窗打破连续性。

一方面，这是一座静默的建筑，不打算引起任何人的注意，但同时人们站在地平线上又能够看见建筑内的灯光，像一座伫立在埃斯特雷马杜拉田野内的海上灯塔。这个立方体结构在不同的楼层为艺术家和公众设置了不同的通道，此外还有行政区、排练室以及顶层的餐馆，使其有利于成为城市的参考点。与建筑的其他区域相比，该立方体结构通过增加其功能灵活性的独特优势，可安排不同的活动日程。

另一方面，通往广场的斜坡的周围有一个"裂缝"式区域，其内的入口大厅是一处多功能空间，与室外流畅地衔接起来。这种布局可用作多种用途，如临时展厅。支撑展览空间的屋顶的柱子也可以作为支撑展品的柱子，从而将结构变成家具。一处扩建的、流动且连续的公共空间的末端为礼堂，礼堂通过服务区与和设备服务后台来与对面连接，分为三层，不但在地下围合出环形空间，并且还成为礼堂的服务空间。

该建筑规划也与周围环境的气候条件紧紧联系在一起，在冬季和夏季，建筑利用混凝土墙体和绿色覆层来维持相对恒定的、舒适的温度，从而形成良好的热惰性。

围绕展览空间设置的整个游览路线全部由白色的光纤玻璃和硅树脂镜片进行装饰，这些镜片或拉伸，或扭曲，将空间在视觉上进行了延伸，以展现虚幻的情境。

礼堂采用的材料为绿色聚碳酸酯板，配上非直射光产生的效果，板材使整座礼堂变成没有明确维度的水世界。空间每一侧的布局都没有明确的材料界限，混凝土的颜色和周围土地的颜色相同，植被的颜色又和线状物（立面的绳子和产生室内饰面如水般特质的材料）融为一体。这里的氛围产生了微妙的变化，空间流转，从白天到黑夜，从东侧到西侧，从自然天成到巧夺天工。

Vegas Altas Congress Center and Auditorium

Vegas Altas Congress Center and Auditorium grows in an ambiguous peripheral location, in a land that is both an urban and an agricultural boundary. The architectural proposal is intended to highlight this timeless condition of a building belonging to the Vegas as a free-standing building, floating in the countryside like a giant bale of straw with a flat horizon, free and fertile. This intervention gives an ample public green space to the town with a green cover over most of the program and exhibition halls. The cube wrapped in a woven web of ropes is the only building perceived above ground, appropriate for both the visibility required and also to absorb the large volume needed for a stage box. The strips of vegetation with herbaceous straw-consistency plantation and its surface ripped with skylights that lights the underground spaces constitute the true facade of the building.

On one hand, it is a silent object that aims to go unnoticed, but at the same time lights are visible on the horizon, like a lighthouse in the sea of the Extremadura's field. On the different floors of the cube stands different accesses for both artists and the general public, administration, rehearsal rooms and a restaurant located on the top floor, a vantage way to become a city reference. It is intended

that the cube accommodates the uses capable of being used with a different schedule than the rest of the building, taking advantage of their independence to multiply its programmatic flexibility.

On the other hand, a crack opens around a ramp access square. The entrance hall is a multipurpose space in fluid connection with the outside. Its layout can be used for multiple purposes, such as temporary exhibition hall. The pillars supporting the roof of this architectural exhibition space can function as support for exhibitions, transforming its structure into furniture. An expanded liquid and continuous public space with the auditoriums at its ends, which bind on the opposite side by the area of services and facilities serving backstage, is divided into three levels closing a ring under the ground and serving both auditoriums.

The architecture proposal adheres to the climatic conditions of the environment. It has a high thermal inertia through the concrete walls and green cover to maintain a constant comfortable temperature in winter and summer.

The entire tour around the exhibition space is accompanied by a white, tense and warped mirror sheet of fiberglass and silicone, which extends the space reflecting an unreal environment. Auditoriums are made with green coloured polycarbonate that by the effect of indirect lighting becomes a water world without a precise dimension. On either side, there is a material ambiguity in the general configuration of the building, in the tone of the concrete that is the same of the surrounding land, in the colours of vegetation transferred to the threads that make up the ropes of the facades and the watery nature of the interior finishes. There is also a fluctuating atmosphere where spaces change its character from daylight to night, from east to west, from natural to artificial.

项目名称：Vegas Altas Congress Center and Auditorium
地点：Villanueva de la Serena, Badajoz, Spain
建筑师：Luis Pancorbo, José de Villar, Carlos Chacón, Inés Martín Robles
工料测量师：adobearquitectura_Manuel Trenado, José Luis Gomez
结构工程师：Mecanismo_Juan Rey, Pablo Vegas, Jacinto Ruiz Carmona
设备工程师：Úrculo Ingenieros_Rafael Úrculo, Sergio Rodriguez
音效师：Arau Acustics_Higini Arau
模型制作：Gilberto Ruiz
施工单位：Placonsa_Eloy Montero, Julio Oreja, Site Manager
甲方：Junta de Extremadura
用地面积：16,470m² / 总建筑面积：3,557.70m² / 有效楼层面积：6,873.26m²
竞赛时间：2008 / 施工时间：2010.1—2014.9
摄影师：©Jesús Granada

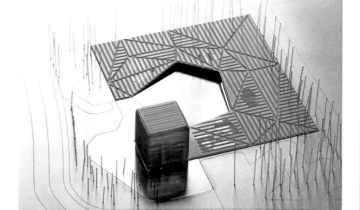

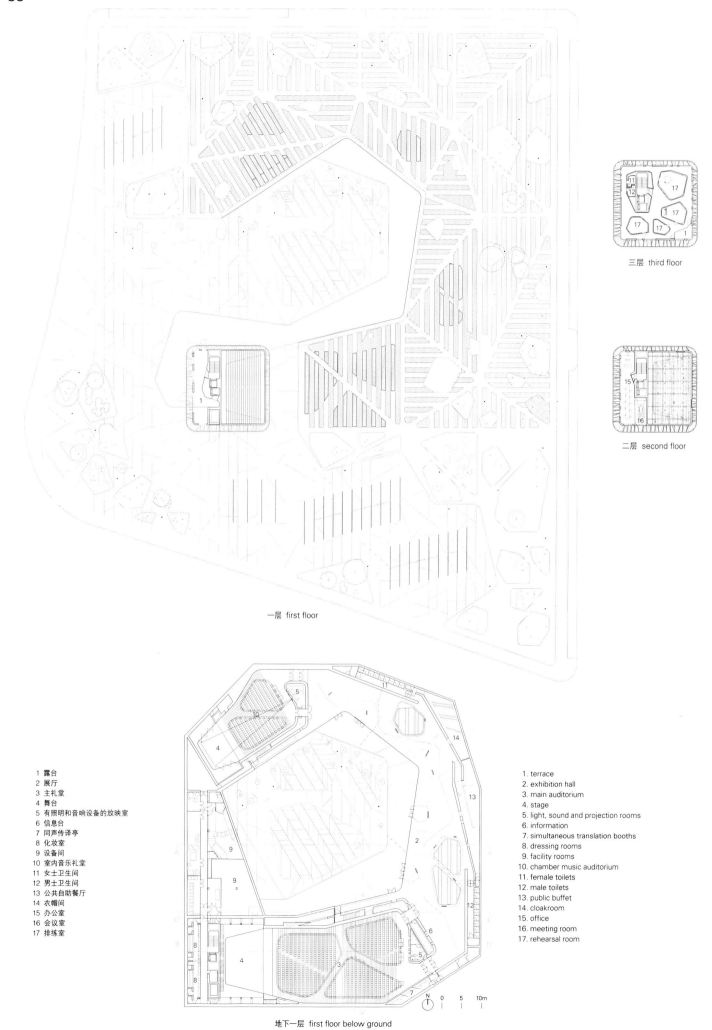

三层 third floor

二层 second floor

一层 first floor

地下一层 first floor below ground

1 露台
2 展厅
3 主礼堂
4 舞台
5 有照明和音响设备的放映室
6 信息台
7 同声传译亭
8 化妆室
9 设备间
10 室内音乐礼堂
11 女士卫生间
12 男士卫生间
13 公共自助餐厅
14 衣帽间
15 办公室
16 会议室
17 排练室

1. terrace
2. exhibition hall
3. main auditorium
4. stage
5. light, sound and projection rooms
6. information
7. simultaneous translation booths
8. dressing rooms
9. facility rooms
10. chamber music auditorium
11. female toilets
12. male toilets
13. public buffet
14. cloakroom
15. office
16. meeting room
17. rehearsal room

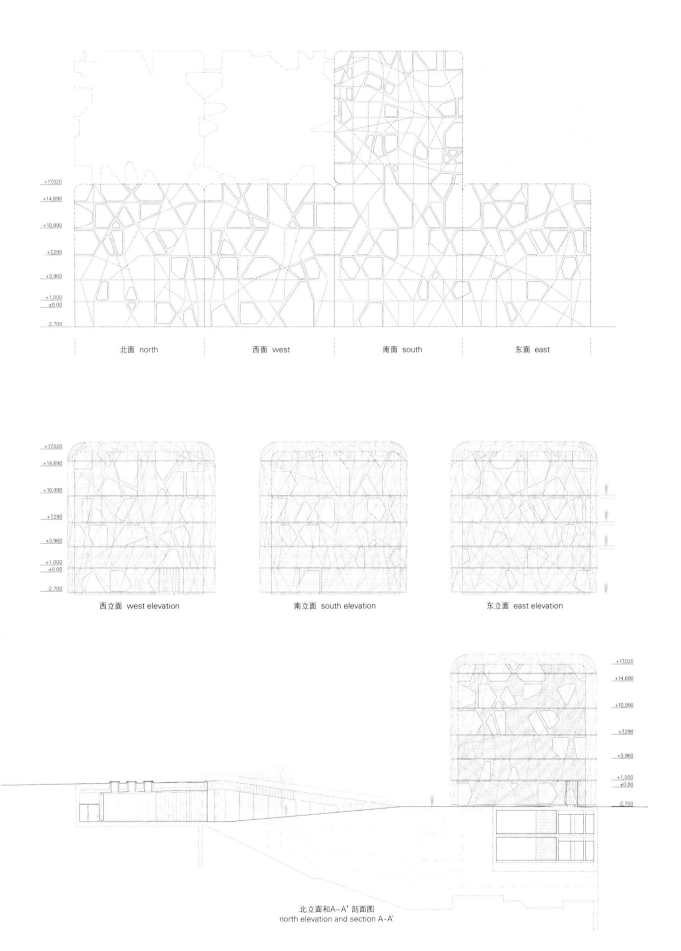

北立面和A-A'剖面图
north elevation and section A-A'

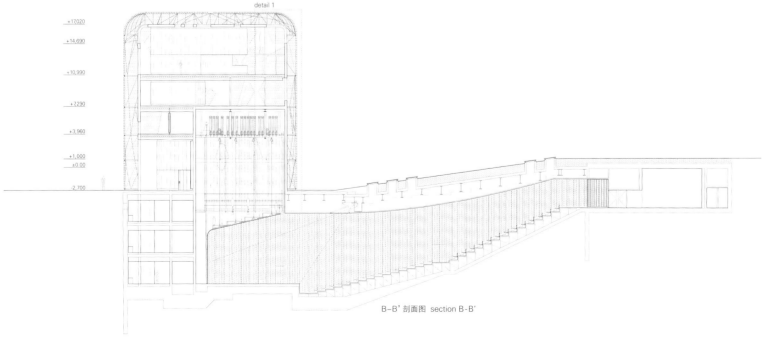

B-B'剖面图 section B-B'

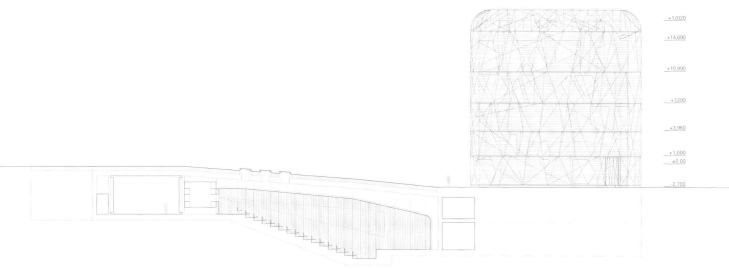

C-C' 剖面图 section C-C'

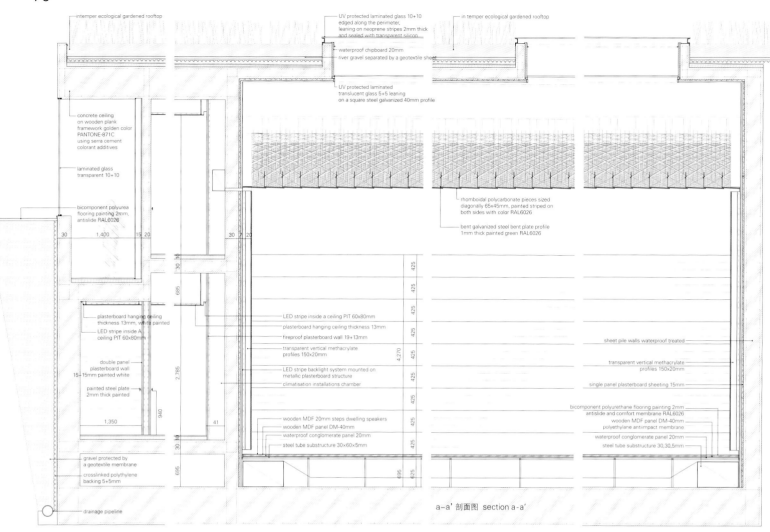

a-a' 剖面图 section a-a'

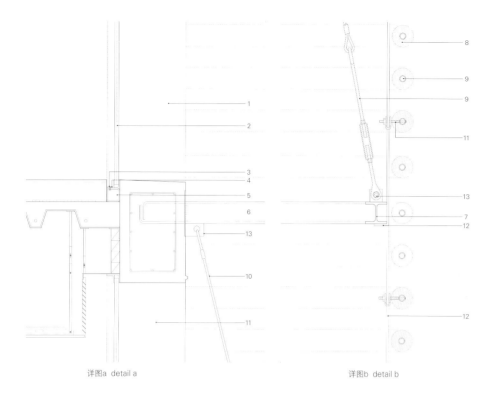

详图a detail a   详图b detail b

1. hollows edges in golden concrete
2. fixed window, glass with curved edge laminar, type stadip 10+10mm
3. galvanized steel profile 35.3 L-painted
4. galvanized steel curved profile 60,40,5
5. window frame of galvanized steel support of perimeter 60,4
6. cantilever support of the external facade, HEB-100 miniated and painted
7. perimeter profile HEB-100, miniated and painted
8. 12 braided ropes polypropylene rope diameter 12cm
9. interior galvanized steel tube
10. bracing of cantilevers, galvanized braided cables
11. anchor of the rope, mechanized bar galvanized with nut
12. anchor plates for fixing of cantilever tensors, miniated
13. painted steel plate

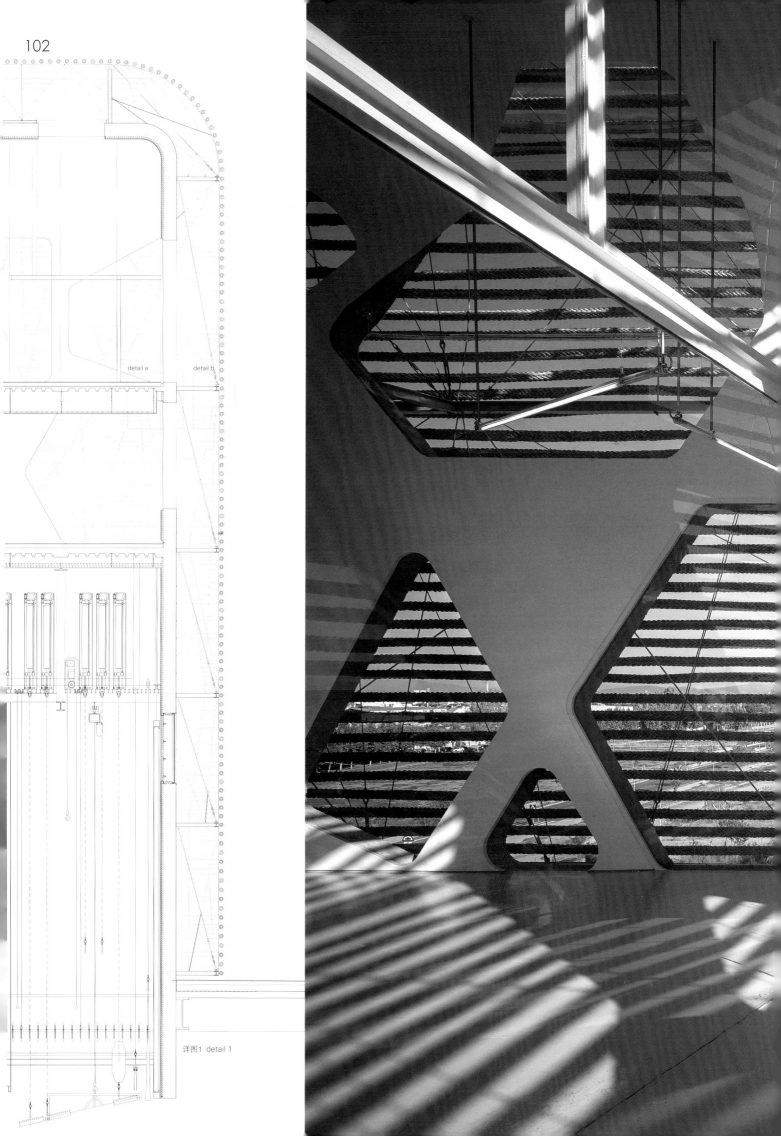

详图1 detail 1

办公室景观
Changing Landscape

现如今，充满积极氛围的办公室的室内多追求个人主义、开放性、以及空间的专属性。工作环境是随着商业、行业——即所谓的供求变化而变化的，当今，我们不难看到一些开放的平面、公司或产品、以及采用智能设计且能够互动的办公室和工作场所。来到一间创新的、有趣的、明亮的、智能的、整洁有序的办公室，对于员工、领导、客户、供应商、保洁人员、迎宾人员、或是巡查人员来说都是一种令人兴奋且难忘的经历。设计有可能对工作环境、声誉以及生产力产生创造性或是破坏性的影响。而处在办公室内的人们的幸福感与持续的努力也会自然而然地发展。整体水平的实际场景都会产生一种有趣的、富有成效的结果。正式与非正式的办公室环境的并置、冒险的空间地形、以及空间内的空间、小屋、舱式空间、袋形区、盒式空间，甚至午睡区正变得越来越普遍。

Today's positive office interior strives for individualism, openness and spatial inclusivity. Workplace environments have changed along with the evolution of business and industry – demand and supply. Today we see open plan, company and product, cleverly designed and communicated office and workspaces. Arriving at an innovative, fun, bright, smart and well-organised office as an employee, director, client, supplier, cleaner, caterer or courier should be an exciting and memorable experience.

Design can make or break the workplace environment, reputation and productivity. Wellbeing and consequent effort of the people occupying the office space go naturally hand in hand. Practical scenarios on a holistic level will lead to an interesting and productive space. Juxtapositions between informal and formal office environments, adventurous spatial topography and spaces within spaces; hubs, pods, pockets, boxes and even sleeping or power napping areas are becoming more and more common.

空中食宿公司都柏林办公室_Airbnb Dublin Office/Heneghan Peng Architects
SoundCloud总部办公室_Soundcloud Headquarters/KINZO
Kashiwanoha开放创意实验室_Kashiwanoha Open Innovation Lab/Naruse Inokuma Architects
Crosswater办公室_Crosswater's Office/Threefold Architects
马德里马塔德罗文化工厂_Cultural Factory in Matadero Madrid/Office for Strategic Spaces
IBC创意工厂_IBC Innovation Factory/Schmidt Hammer Lassen Architects
Lowe Campbell Ewald总部_Lowe Campbell Ewald Headquarters/Neumann/Smith Architecture
1305工作室_1305 Studio Office/1305 Studio
CDLE办公室_CDLE Office/R-Zero Arquitectos
办公室景观_The Changing Landscape of the Office Interior/Heidi Saarinen

of the Office Interior

在开启富有特色的办公室之旅前,我们先看一下这个话题:办公室的内部环境。当工作环境成为我们从事专业工作的地方,并且也是个人和创造性人员对行业起到绝对性作用的地方,那么这处办公区和工作空间便是成功的。从办公室内部的环境、资源,以及相关活动中汲取灵感,能帮助我们实现生活和工作的整体平衡。然而,没有人的办公室不能称其为办公室,这对日常活动、日常事务、解决具体的行业难题和要求具有关键的作用。

回顾以前,我们从黑白电影中可能会观察到这样的场景,办公室内的道具,如胶木材质的电话、打字机、电报机、传真机,以及后来出现的大型电脑显示器、制图板和桌子上的小装置,这些呈现出来的就是我们所能想象的过去的办公室场景。这些情节经常出现在早期电影呈现的办公室中,主要描绘传统的、兼容的办公室环境。[1] 一个看起来很能干的秘书,要么正在拼命地敲击打字机,要么就是极力阻止不被接见的访客进入领导办公室。这种分等级的环境和组织模式如今已经被现代的、充满活力的多样环境所代替,在这里每个人都扮演着一个重要的角色。

进入办公室我们不再需要在考勤簿上盖章。灵活自由的办公时间、轮用办公桌,以及家庭办公成为在办公室之外执行专业任务时普遍接受的模式,创造性的解决方案甚至会产生在会议室以外的地方,而且也会考虑到创造与创新之间的一种健康的平衡。

在当今的数字时代,不仅仅是工具和设备改变了外观和应用。人也一样,各色各样,有着不同的背景和经历,不同的想法和诉求,所有的一切都需要可选择的空间来形成创新的思维和集体创新。无论私人空间还是公共空间都必须充分考虑其功能性空间的建造。这处空间可能以会议室、研讨会或是娱乐区的形式存在。社交空间已经成为许多办公室的核心,在那里,同事们可以尽情放松,通过自由闲谈来持续地创作,也会在所谓的办公室操场打几场乒乓球。

Before embarking on the journey through the featured office spaces, we will be briefly looking at the topic, the Office Interior. Having work environments within which we can thrive not only professionally, but also individuals and creative human beings are absolute key for a business, office and workspace can be successful. Drawing inspiration from the surrounding spaces, resources and activities within an office interior will help achieve overall life/work balance. However, an office cannot be complete without the people, which is key to the day-to-day activities, routines and industry specific intricacies and demands.

Glancing back in time, we may envisage a scene from a black and white film, where office props like Bakelite phones, typewriters, telex and fax machines and later large computer monitors, drawing boards and desk gadgets are what we may think of as the office of the past. These scenarios were often set within office environments as seen in the early films portraying a stereotypical, non-inclusive office environment[1]. An efficient looking secretary may be seen busily clicking away on a typewriter or stopping unwanted visitors from entering the manager's office. This type of hierarchical environment and set-up has been replaced by the contemporary, vibrant and diverse environment of today where everyone has an important role to play.

We no longer need to stamp a clocking-in card at arrival in the office, as flexible working hours, hot-desking and home working are widely acceptable modes of carrying out professional duties alongside the office environment. Creative solutions reach outside the boardroom and allow for a healthy balance between create-innovate.

In today's digital age, it is not only the tool and equipment that have changed the appearance and use. There is a diversity of people from different backgrounds and experiences, with different ideas and voices, all requiring eclectic spaces for forming creative ideas and collective innovation. Private and public space must also be taken fully into consideration for a functioning space. This may be in the form of meeting, workshop and recreational zones. The social space has become central to many offices, where colleagues relax and continue creations through informal discussion and in the "office playground", whilst playing table tennis.

线图提供:©Heneghan Peng Architects

Kashiwanoha开放创意实验室,千叶市,日本
Kashiwanoha Open Innovation Lab in Chiba, Japan

Kashiwanoha开放创意实验室,千叶市,日本
Kashiwanoha Open Innovation Lab in Chiba, Japan

在日本千叶,由成濑·猪熊建筑设计事务所设计的的Kashiwanoha开放创意实验室的外观的技术含量并不高,且其表面未加修饰,饰面保持天然的状态,但是高技术的创意工作填补了其核心的地位。暴露的表面充当了颜色和质地的背景。多种功能在内部和谐地交织在一起,利用高度和环境上的改变来对这座创意中心进行改造,新理念和新企业的启动都会有助于当地经济气候的萌芽。

社交场地、会议室、安静的工作场所大多散布于开阔的工业空间周围。大规模的墙面制图向社交环境致敬。精心考虑的细节增加了室内的趣味性,例如带有奇怪图案的座垫、精心布置在工作空间周围的草木。企业家、个体、公司共同努力创造产品,形成理念,为行业、人们以及投资提供机会。

当你进入KINZO设计的柏林网上社交音乐平台SoundCloud总部时,首先映入眼帘的是一个操场,在它的设计中,每一个细节都能被充分地考虑到,工作、生活和休闲达到了一种很酷的平衡。

该设计是一个开放的平面,空间内部亦有空间,安静的私人区与公共空间结邻为伴,在这些地方,共享和友好相处才是关键。在一个寒冷的冬日午后,悠闲地坐在火炉旁,感受唱片机带来的丝丝寒意,使人想到一个假日小屋。坐卧两用的沙发是午睡的绝佳之地,甚至令人羡慕。

建筑师保留了仓库改建的工业特征,与附近的老柏林墙的地点相呼应,与此同时还要考虑通过简单且有效的工作、概念、社交和网络空间来营造明亮且全新的氛围。声乐材料和照明设备反映了办公室的音频主题。裸露的横梁和砌砖、开敞式的工业楼梯,以及木材围成的封闭空间位于这一开放的景观内,使其本身成为一处折中的多功能空间。

位于中国上海的1305工作室采用了盒中盒的设计方案,所有的设计理念和创意活动都在主结构内形成和举办,此外,这里还设有其他盒式结构,如房间的坐椅、套间、阅览和工作室,可服务于各种活动。有趣的

A clever low tech office look with bare surfaces and natural finishes, allows for high tech creative work to fill the focus at Kashiwanoha Open Innovation Lab in Chiba, Japan by Naruse Inokuma Architects. Exposed surfaces act as the backdrop to the color and textures. Various functions occupy the interior harmoniously using heights and environmental approaches to operate this innovation center. New ideas and start-ups are germinating to aid the local economic climate.

Social interaction, meeting, quiet and work areas are scattered generously around this sizable industrial space. Large-scale wall graphics compliment the social space. Carefully considered details add to the interesting interiors, such as odd patterned seat covers and greenery strategically placed around the workspaces. Entrepreneurs, individuals, companies all work together creating products, ideas and opportunities for business, people and investment.

A playground greets you when entering the Berlin HQ of music online platform SoundCloud by KINZO. Everything has clearly been considered; there is a cool balance between work/life/play.

The design is an open plan, with spaces within spaces, private and quiet zones neighbouring happily with the communal areas, where sharing and friendships appear to be key. Taking breaks sitting around the wood burner on a cold winter afternoon, chilling to some vinyl tunes from the record player, may remind one of a holiday chalet. The daybeds, a favourite, must be the envy of many, being a wonderful space for a power nap.

The architects have retained the industrial features of the converted warehouse, echoing the location near the old Berlin wall, and at the same time allowing for a bright, new ambience to become evident through the simple yet effective work, concept, social and networking spaces. Acoustic materials and lighting reflect the audio theme of the office. Exposed beams and brickwork, open industrial staircases and enclosed timber spaces amongst the open plan landscape make this an eclectic and multifunctional space.

Boxes within boxes is what 1305 Studio Office interior in Shanghai, China is designed around, and the ideas and creative activities all take place in the main container like space with other box structures – seating, rooms within a room, reading and work spaces

Lowe Campbell Ewald总部，底特律，密歇根州，美国
Lowe Campbell Ewald Headquarters in Detroit, Michigan, USA

概念形成设计的基础；1305工作室的建筑师称这个空间充满了"无限的可能性"。不同层次的空间变化使办公环境和活动也有所变化，它们设有不同的功能，从讲室和图书馆，到跑道和工作平台。室内的空间组织和物质性使创新精神成了重中之重。人们在建筑的所有地方都能看见当地的遗产，并且望见室外景观以及景观内嵌入结构之外的墙体。

位于底特律，由诺尔曼·史密斯建筑事务所设计的Lowe Campbell Ewald广告公司总部是一处非常独特的场所，因为它的选址和公司精神都独树一帜。精选的低成本材料应用于整个结构。二次使用的隔板在内部形成屏风，完美地隔了私人区和公共区。报废的门经过巧妙切割可以用于构建内部吊舱。许多具有可持续性的特色都被容纳其中，比如回收的谷仓木材、混凝土柜台和织物都被赋予了新的用途。所有的这些细节都赋予空间当前的自由感与充满活力的能量。该机构之前的印刷机板与大堂的天花板相连接，有力地突出了公司的特色。开放的过道和楼梯指引着人们穿过建筑，而中庭内全高的屏风展现了图像和视觉的交流，并且不断地改变着办公室的空间氛围。

在世界范围内，在私人专属空间内出租客房，这一理念与Heneghan Peng建筑师事务所设计的空中食宿公司都柏林办公室的布局相呼应。表面为玻璃的会客舱考虑了可视性，并且引进光线，形成内部整体开放式的平面。贯穿整个室内的几个主题能让人联想到世界范围内各个地方都被公司开发利用。这里设有一间传统的爱尔兰酒吧，十分注重所有的细节参考和关键之处，如马蹄状的柜台。受阿姆斯特丹公寓的启发，中央会客室可以巧妙地分成两个独立的空间。办公室的休息平台，也可以称之为小山，似乎是从办公室的其他区域升起来一样，将静室/阅览区、讲室、储存区融为一体。建筑师将私人区和公共区实现了完美结合。

由丹麦建筑师Schmidt Hammer Lassen建筑师事务所设计的位于科灵的IBC创意工厂坐落于一个翻新的涂料厂内，作为一个创意景观，它

– assigned for the various activities. Interesting concepts inform the basis to the design; the architects, of 1305 Studio, describe it as a space with "many limitless possibilities". The space changes between several levels identifying the office environments and its activities, with name tags from lecture and library, to runway and work levels. It's spatial organization and materiality let the creative spirit remain key. Here and there one can glimpse the local heritage and view the exterior landscape and walls beyond this insertion in the landscape.

Advertising agency Lowe Campbell Ewald's Headquarters in Detroit by Neumann Smith Architecture is a unique place, for a unique location and company ethos. Carefully selected low cost materials are applied throughout the building. Reused pallets form interior screens and sections off private and public areas. Salvaged doors are cleverly cut up and used to build interior pods. Many sustainable features are included, such as reclaimed barn wood, concrete counters and fabrics that have been given to a new use. All these details give the space a current feel of freedom and lively energy. Printing press plates from the agency's past are attached to the ceiling in the lobby space, enhancing identity of the company. Open walkways and stairs guide through the building and a full height screen in the atrium display graphic and visual communication constantly changing the spatial ambience of the office.

The idea of renting rooms in private spaces around the world is echoed in the layout of Airbnb Dublin Office by Heneghan Peng Architects. Glazed meeting pods allow for visibility, and let light in and contribute to an overall open plan of the interior. Several themes are seen throughout, all reminiscent of various locations around the world used by the company. There is a traditional Irish pub, with all the key references and details, such as a counter shaped like a horseshoe. Inspired by an Amsterdam apartment, the central meeting room can ingeniously be divided into two separate spaces. Within the office space a seating platform, or "hill", appears to rise above the rest of the office, incorporating quiet/reading, lecture and storage areas. The architects have combined the private and public in "harmony".

Housed in a refurbished paint factory, the IBC Innovation Factory

IBC创意工厂,柯灵,丹麦
IBC Innovation Factory in Kolding, Denmark

照片提供:©Adam Mork

的内部使用传统的自然形态的材料。在这里人们可以找回用于社交活动的游戏区,就像本文中提到的其他设计方案中呈现的那样。在这里人们可以在学习或讨论之余进行桌球或其他娱乐活动。这种理念旨在尽可能地达到学习、工作和人际交往的时间的自然平衡,为未来的创新者创造条件。感官元素在大天窗的使用中得以体现,透过天窗,室外场景映入眼帘,植物、水声、鸟鸣融入室内。

Strategic Spaces办公室设计的马德里马塔德罗文化工厂被建筑师看作文化产业的孵化器。该设计团队致力于以低成本、高创新水平和高速度来完成目标。

标准尺寸的当地松木应用于许多主要的空间,比如楼梯、大型聚碳酸酯板墙壁系统一侧的平台,墙壁系统能迅速地安装起来,进而对空间进行改造。室内的材料和构造应用于不同的布局中。建筑师把这些空间描述为"浓缩于全面,致力于宁静"。设计者利用该建造过程来探索如何将设计方案转变为可持续发展的工具。

由Threefold建筑师事务所设计的Crosswater办公室,其核心是木质桥体结构,总跨度为64m。它的中心部分连接着会客室、会议室,并且鼓励设置社交场所和创意角落。这里有通道、楼梯,以及沿着主路下至桥底的聚会点。预制的交叉层木制成的桥体与水平面和垂直面相互呼应,而覆盖在墙壁和天花板的肋架也同时将不雅观的服务管道和音效设备遮盖起来。周围的景观在桥上一览无余。照明系统融入木质天花板内,悬浮的荧光灯管照亮部分公共区域,淡淡的色彩有助于指引办公室空间。工作人员和游客混杂在咖啡馆、酒吧以及共享的办公空间的露台和陈列室内。

墨西哥的R-Zero建筑师事务所设计的CDLE办公室是通过"创造而非建造"的理念设计的办公空间。这座建筑是受政府保护的,所以必须尽力保持其特性和历史的完整性。原表面、裸露的砖墙,以及建筑的保护措施一直是设计过程中的关键。修复的门和窗框、天花板以及当代照明系统的细节加固了原始的肌理和构造。设计方法充分考虑了历史回

in Kolding, Denmark by Schmidt Hammer Lassen Architects is fitted out as an innovative landscape, using traditional and found materials for its interiors. Here the playful areas of social interactions return, as seen in many of the other schemes covered here. Table sports and other recreational activities can take place in-between learning and discussion. The idea is to make a balance between work, learning and social time as natural as possible, for the innovators of the future. Sensory elements have been incorporated through the use of large skylights, inviting the outside in. Plants and the sound of water and birdsong are also parts of this interior.

Cultural Factory in Matadero Madrid by Office for Strategic Spaces is referred to by the architects as an incubator of cultural industries. The design team worked to a low budget brief but kept levels of innovation and pace high.

Standard sized local pine was used for many of the main constructed spaces, such as staircases and platforms alongside large polycarbonate sheet wall systems that were quickly erected to totally transform the space. Materials and construction of the interiors were used in many different configurations. The architects describe these spaces as "compact to expansive, busy to silent". The designers explored through the process how the design programme could allow change to act as "a tool for sustainability".

Crosswater's Office by Threefold Architects is dominated by its wooden "bridge" structure, spanning 64m. This central component attaches to meeting pods, conference rooms, and encourages social areas and creative corners. There are walkways, stairs and meeting points along the main path off the bridge. The prefabricated cross-laminated timber bridge echoes the vertical and horizontal planes and fins that cover the walls and ceilings and at the same time, hide unsightly service ducts and aid acoustics. Views of the surrounding landscape can be seen from the bridge. Lighting is incorporated into the timber ceilings, suspended fluorescent light tubes light up parts of the communal areas and hints of colour help guide through the office space. Staff and visitors mingle in the cafe, bar and terrace of the shared office space and showroom.

CDLE Office in Mexico City by R-Zero Arquitectos has been mind-

马塔德罗文化工厂，马德里，西班牙
Cultural Factory in Matadero Madrid, Spain

忆，完美地保留材料和质地，与过去相呼应。这个地方曾经被用做住房、面包店、妓院，之后这些印迹也随着时间慢慢消失了，这个地方上演了很多故事。今天这座建筑被重新设计为三个客户的复杂的工作空间，仍然述说着它的历史和建筑过往。

本章内所讲述的空间在当今的办公室和室内工作区中都占有一定的地位。从传统的意义上讲，一些办公室未必可以简单地称为办公室，但可以说更像车间、学习空间、小径，或是公共会客区，就像在Crosswater新办公室看到的桥一样。甚至还有一些这样的空间，如临时办公室，或许是临时的一个生活据点，同样也需要灵活的办公室装修。还有主题办公室，能够将产品、商业以及故事有趣地整合在一起，如空中食宿公司都柏林办公室。

现如今，办公室的整体感觉是随着产业的发展而变化，也随着食物链上每个参与者的供求变化而变化，同时需要考虑怎样创造出最好的环境，使其既有竞争力，又空间宽敞，广纳建议并且具有发展空间，就像柏林新SoundCloud总部和马德里马塔德罗文化工厂那样。本文所叙述的几个办公室都可以被看作是画布，其中所涉及的关键材料、颜色、照明以及总体构造和设计是可以共存的，各个元素发挥所能，变得与众不同。这样来看，人类追求的形态就是，不仅要通过人机工程学和设计，也要直接通过理念，以及在那些各式各样的、令人振奋的办公室空间进行创意工作的人们来创造空间。

fully designed as office space by "creating not constructing". This building is government protected and must keep its characteristics and history as intact as possible. A celebration of raw surfaces, exposed brick walls and architectural conservation has been absolute key in the design process. Restored door and window frames, ceilings and contemporary lighting details underpin the original fabric and construction. The design methodology takes into account memory and echoes from the past, respectfully retaining materials and textures. Having served as housing, bakery and brothel and later neglected, this space tells many stories. Today, redesigned into sophisticated workspaces for three clients, the building is still clearly narrating its history and architectural past.

The spaces covered here, all take their own stance on the contemporary office and work place interior. Some "offices" are not necessarily simply offices in the traditional sense, but may be more about workshops, learning spaces, paths and communal meeting areas as seen in for example the Crosswater's New Office, the Bridge. There may even be spaces that house offices temporarily, or short life locations, requiring an equally flexible office fit out. Themed offices, playfully integrates product, business and narrative as in the Airbnb Dublin Office.

The sense of the office today is very much about the speed at which business moves, the supply and demand of each player in the food chain, and how best to create environments that are competitive yet spatially generous and open to suggestions and development as in the New Sound Cloud Headquarters in Berlin and Cultural Factory in Matadero Madrid. Many offices covered in this article can be seen as the canvas within which key elements of material, colour palettes, lighting and overall construction and design can coexist, making these spaces special in their own right. It does appear that the human form is what creates the space not just through ergonomics and design but also directly through the ideas and the creative beings that operate within these different and exciting office spaces. Heidi Saarinen

1. *Man in a White Suit*, Mackendrick, A., 1951.

空中食宿公司都柏林办公室

Heneghan Peng Architects

2014年2月,空中食宿公司将其位于都柏林的办公室搬至Watermarque大楼。有团队专门负责整个空中食宿的流程,从预定到入住和整体体验,因此都柏林分公司监管大部分空中食宿公司的户主和租客的体验。办公室由都柏林当地的建筑公司Heneghan Peng设计,设计要求建造一系列富有创意的合作空间,这些空间能够沿袭公司的精神和文化。

和空中食宿公司的旧金山总部一样,都柏林办公室的设计被要求引进全世界空中食宿公司的租赁房屋中的一些有趣元素,这些元素都在开放式办公室四周的个人空间得到应用。空中食宿公司的总裁Brain Chesky要求,员工和来宾能够一直看到从入口区域下至后面的阶梯处的风景。因此,每间会客室的两侧都有玻璃墙,景色一览无余。

这些受到启发的空间不仅提供了单独工作和会客的区域,它们还被用作非正式的休闲区域,且每个房间内还提供了设有座位的壁龛,位于一侧的室内墙体内。这些空间的室内设计也参考了同样的元素。

仿照一间阿姆斯特丹公寓而设计的会客空间由两个相连的单间组成,这两个单间被隔成独立的空间,显示了Heneghan Peng设计的多功能性。接待处被改造成爱尔兰酒吧,带有一个马蹄状的吧台,这也是参考了都柏林的办公室的设计。此外,地面采用了三种不同风格的瓷砖,它们都来自许多不同年代的酒吧,还有老式爱尔兰电话亭。

办公室里还有一个长椅,它是从Heneghan Peng2012年参加威尼斯国际艺术双年展时设计的爱尔兰亭子中拆下来的。长椅长12m,由6个内部相连的部分组成,带有6个可旋转支点和5个转换轴,主要是为了在Artiglierie展区后部提供休息场地。长椅不坐人的时候保持水平状态,坐上去类似于跷跷板,一边高一边低,使用者之间可以进行互动。

该建筑后方的一片空间是黑暗的大角落,没有自然光,没有窗户,于是建筑师被要求改造这个区间。他们选择建造一处阶梯,用作开会、休闲以及非正式的办公区域。在那里员工可以俯瞰整片办公区。

Airbnb Dublin Office

Airbnb moved its Dublin office into the Watermarque building in February 2014. The Dublin based arm of the company oversees much of the host and guest experience for Airbnb, with teams dedicated to the entire Airbnb process, from booking through to staying and overall experience. Designed by Heneghan Peng, a local Dublin based architecture firm, the brief was to create a series of creative and collaborative spaces which emulate the Airbnb ethos and culture.

项目名称:Airbnb Dublin Office
地点:The Watermarque Building, Bridge Street & South Lotts Road, Dublin 4
建筑师:Heneghan Peng Architects
机械&电气工程师:Axis Eng
项目经理:KMCS, Acoustic
顾问:AWN
用地面积:3,200m² / 总建筑面积:10,000m²
施工时间:2013—2015 / 竣工时间:2015
摄影师:©Ed Reeve(courtesy of the architect)

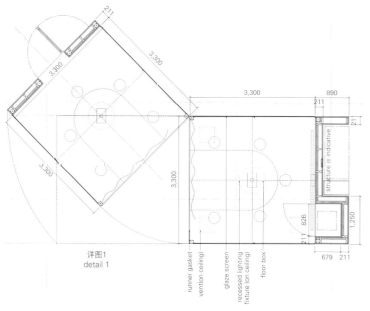

详图1
detail 1

runner gasket
vent(on ceiling)
glaze screen
recessed lighting fixture (on ceiling)
floor box

As with Airbnb's San Francisco HQ, the architects were also asked to introduce some of the most interesting Airbnb listings from around the world into the office design. These are channeled into individual pods dotted around the open plan office. As these were implemented, Airbnb's CEO Brian Chesky requested staff and visitors to the office to have a continuous view of the space – from the entrance area down to the back end stepped area. This led to each meeting room having window walls on two sides, allowing for uninterrupted views.

The inspired spaces not only provide areas for solitary working or meetings, they can also be used as informal relaxing areas with each housing a seated alcove space on one of the internal walls. The design for these is referenced from the same listing as the interior.

Showcasing the versatility in Heneghan Peng's design, the meeting space modeled on an Amsterdam apartment is mirrored in two connecting pods which can be split into two separate spaces.

There are also references to Dublin with the reception area modeled on an Irish pub with a horseshoe shaped bar(the floor uses three different styles of tile found in pubs from different eras) and traditional Irish telephone boxes.

The office also houses a bench from the Pavilion of Ireland's 2012 contribution to the Venice Biennale. Designed by Heneghan Peng, the long 12 meter bench is constructed of 6 interlinked sections, 6 rotation-only fulcrums & 5 translation pivots, and was designed to provide a place to sit at the end of the Artiglierie. The bench, when at rest in balanced equilibrium is horizontal. As users sit on the bench the height of each section alters, creating dips and peaks for user interaction.

The architects were also charged with providing a resolution for the back end of the office which features no access to natural light or windows. Here they chose to create a stepped area, which is now used as a meeting space and relaxed and informal working area, allowing staff to look over the office.

1. open office
2. welcome area
3. drinks & beverage
4. eating & collaboration
5. kitchen
6. servery
7. break space
8. meeting room - tree house
9. meeting room - terminal 1 & 2
10. meeting room - Beijing
11. meeting room - Madina & Marrakech
12. meeting room - St. Martin
13. library
14. IDF room
15. general storage

1 开放的办公室
2 迎宾区
3 饮品区
4 餐饮&协作区
5 厨房
6 服务区
7 休息空间
8 会客室-茶室
9 会客室-终端区1&2
10 会客室-北京
11 会客室-Madina& Marrakech
12 会客室-St. Martin
13 图书馆
14 IDF室
15 总储存室

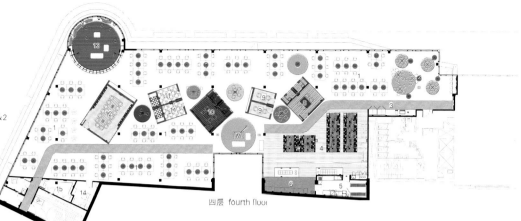

四层 fourth floor

1. open office
2. eating & collaboration area
3. training room1
4. training room2
5. IT help desk
6. IT store
7. MDF room
8. meeting room - Beat Suit
9. phone booth
10. meeting room - captain's cabin & livingstone
11. meeting room - Tokyo
12. meditation & Yoga-zen area
13. project room
14. general storage

1 开放的办公室
2 餐饮&协作区
3 培训室1
4 培训室2
5 IT咨询台
6 IT商店
7 MDF室
8 会客室-Beat Suit
9 电话亭
10 会客室-船长的小屋&活石
11 会客室-东京
12 静思室&瑜伽-禅宗区
13 项目室
14 总储存室

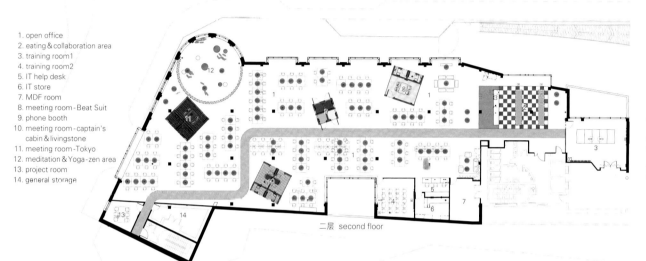

二层 second floor

1. reception
2. eating & front bar
3. eating & meet bay1
4. eating & meet bay2
5. breakfast bar wall
6. kitchen
7. open office
8. video conferencing room - Amsterdam
9. meeting room - Austion
10. meeting room - Lyon
11. meeting room - New York
12. meeting room - Mexico
13. stepped seating - the hill

1 接待处
2 餐饮&前台
3 餐饮-会客区1
4 餐饮-会客区2
5 早餐吧台墙
6 厨房
7 开放的办公室
8 视频会议室-阿姆斯特丹
9 会客室-奥斯汀
10 会客室-里昂
11 会客室-纽约
12 会客室-墨西哥
13 阶梯座位-小山

一层 first floor

1. eating & meet bay
2. video conferencing room - Amsterdam
3. meeting room - Austion
4. open office
5. meeting room - Lyon
6. meeting - Mexico
7. meeting room - New York
8. stepped seating - the hill
9. meeting room - Tokyo
10. phone booth - Nintendo
11. meeting room - Beat Suit
12. eating & collaboration area
13. training room
14. library
15. meeting room - Madina & Marrakech
16. meeting room - Beijing
17. meeting room - terminal 1 & 2
18. meeting room - tree house
19. welcome area

1 餐饮&会客区
2 视频会议室-阿姆斯特丹
3 会客室-奥斯汀
4 开放的办公室
5 会客室-里昂
6 会客室-墨西哥
7 会客室-纽约
8 阶梯座位-小山
9 会客室-东京
10 电话亭-任天堂
11 会客室-Beat Suit
12 餐饮&协作区
13 培训室
14 图书馆
15 会客室-Madina&Marrakech
16 会客室-北京
17 会客室-终端区1&2
18 会客室-树屋
19 迎宾区

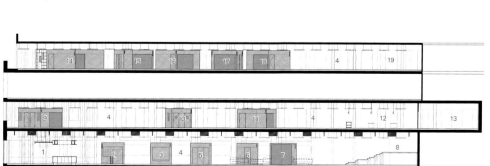

A-A' 剖面图 section A-A'

SoundCloud总部办公室

KINZO

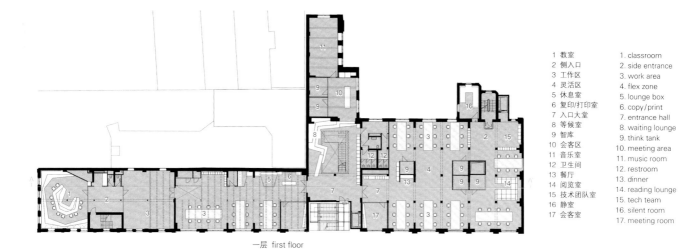

一层 first floor

1 教室	1. classroom
2 侧入口	2. side entrance
3 工作区	3. work area
4 灵活区	4. flex zone
5 休息室	5. lounge box
6 复印/打印室	6. copy/print
7 入口大堂	7. entrance hall
8 等候室	8. waiting lounge
9 智库	9. think tank
10 会客区	10. meeting area
11 音乐室	11. music room
12 卫生间	12. restroom
13 餐厅	13. dinner
14 阅览室	14. reading lounge
15 技术团队室	15. tech team
16 静室	16. silent room
17 会客室	17. meeting room

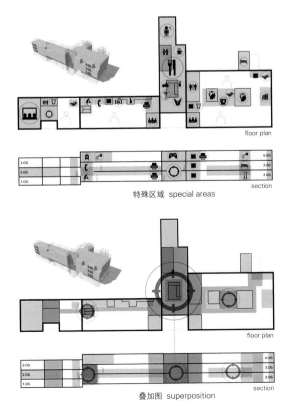

特殊区域 special areas

叠加图 superposition

国际知名的SoundCloud在线平台的新总部办公室占据了大约三层楼的垂直空间。该建筑由前柏林墙附近一个占地约4000m²的老啤酒厂改建而成。设计旨在将这座综合设施打造成为这个前途无量的新兴公司的未来孵化器。受雇于这个主流网上音乐共享互动平台的180位柏林员工未来将在这个一站式办公室里工作。

考虑到SoundCloud平台的迅速发展，为了适应未来扩张的需要，宽敞的办公室设计最多能够容纳350张桌子。

为了给公司和新总部打造出与该平台创新的商业模式和工作结构相匹配的空间个性及建筑框架，SoundCloud公司请KINZO工作室将其新总部设计成为一处鼓励革新和创意的空间。该平台在其柏林办公室景观内被构想为一个受欢迎的灯塔形象，能够在塑造公司形象的同时，也适应SoundCloud未来几年的发展。

在此背景下，KINZO的改造设计利用其对虚拟社区的映射，重新定义了办公室的设计规则。

Soundcloud Headquarters

The new headquarters of internationally renowned online platform SoundCloud cover three levels and approx. 4,000 square metres of an old brewery building close to the former Berlin Wall. The complex, designed to serve as a future incubator for up-and-com-

项目名称：SoundCloud Headquarters / 地点：Berlin, Germany
建筑师：KINZO
工程监理：Karim El Ishmawi, Martin Jacobs / 项目首席建筑师：Timo Nerger
项目团队：Stefanie Pesel, David Schumm, Susanne Trumpf, Laura Liberal, Ina Podzimek, Manfred Kaiser
声学规划师：Akustikbüro Rahe + Kraft GmbH / 照明规划师：Dinnebier + Bheske
技术构建服务人员：Häfner Ingenieure GmbH / 电气工程师：Elaro GmbH
有效楼层面积：4,000m² / 造价：EUR 3.2 million / 竣工时间：2014
摄影师：©Werner Huthmacher

ing start-ups, will become the one-stop office for all 180 Berlin employees of the foremost hub for musical exchange on the net. Anticipating future expansion of the rapidly growing SoundCloud platform, the spacious offices could potentially accommodate up to 350 desks.

To provide the company and its new HQ with a matching spatial identity and the right architectural framework for the platform's novel business and working structures, SoundCloud asked KINZO to turn its new hub into a space that encourages innovation and creativity. The platform envisaged a welcome beacon in Berlin's office landscape that both shapes the company and evolves with SoundCloud over the years.

Against this background, KINZO's conversion redefines the rules of office design with its tangible reflection of a virtual community.

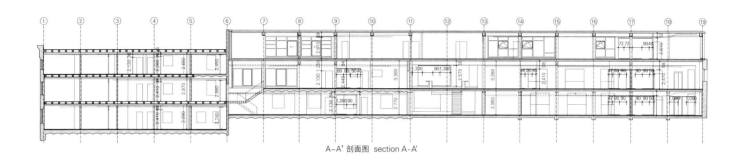

A–A' 剖面图 section A-A'

Kashiwanoha开放创意实验室

Naruse Inokuma Architects

Kashiwanoha开放创意实验室是日本的一个旨在支持企业家创业、促进企业发展和刺激经济活动的创意中心。这个创意实验室就像一个平台，企业和个人在这个平台上共同努力，超越传统的框架，融合思想、技能，并且生产创新产品，提供创新服务，这一切在投资者的支持下利用该体系都得到了便利，进而实现目标。

Kashiwanoha开放创意实验室具有多功能性，为不同领域的顺利沟通提供保证，如就餐、生产、做报告和休息。用户可以在这座综合设施内自由选择各式各样的区域工作，与其他用户共享空间和设施，这样就可以以各种方式和用户接触。该中心好像微型城区，各种活动和事件能同时上演。

在中央公共区，因容纳不同的活动，不同用途的区域相互交义，因此形成了多元区域。这些区域的天花板高度、灯光的色温和内饰也不尽相同，以满足区域功能。这看上去似乎更加约束，限制了空间的特色和用途，但是空间布局有机运转，能满足所有用户的工作风格。20世纪办公室以为管理者提供统一灵活性为基础，而这处空间却恰恰相反，再现了为每名员工设计的灵活性。

在视觉元素方面，设计突出未完工的气氛，目的是让人们的活动成为空间的亮点。公共区天花板上的通风管暴露在外面，可以反射光线。其他构件则采用基础材料如木丝水泥板，增强纤维水泥板，石膏板（用油灰简单抹平，上面刷上透明涂料）来进行装饰。这些细节的设计旨在避免最终的饰面所反映出的限定意义或特定风格。

Kashiwanoha Open Innovation Lab

Kashiwanoha Open Innovation Lab(KOIL) is an innovation center intended to support start-ups of entrepreneurs, promote enterprise developments and stimulate economic activities in Japan. It is a place like a platform where companies and individuals work together beyond a traditional framework and fuse ideas and skills to produce innovative products and services, which are facilitated and realized by the system with investors' supports.

KOIL, therefore, includes various functions for smooth communication crossing over fields, such as having a meal, manufacturing, making presentations and relaxation. Users may choose their place from various locations within the complex freely and work while sharing the spaces and facilities with other users, which allow users to contact with others diversely. The center becomes a place like a miniature of urban city, where various activities and events occur simultaneously.

For the place allowing such activities, diverse spaces have been created with areas that have different applications and intersect

within the central public zone and with various ceiling heights, color temperatures of the lights, and finishes of interior designed to match the areas' functions. It seems to increase restrictions to limit the characters of spaces and usages, however, the layout functioning organically altogether enables the space to meet all kinds of working style of all users. In contrast with offices during the 20th century which were made up based on uniform flexibilities for managers, this space represents the flexibilities for each worker.

About the visual elements, the unfinished atmosphere is emphasized in the design aiming to make people's activities themselves be the charm of the space. The ceiling of share area has exposed duct pipes reflecting the light. The other elements are simply finished using the base materials such as wood wool cement boards, fiber reinforced cement boards, and plaster board leveled with putty covered with clear paint. Through such details, the design intends to avoid restricted meanings or a particular style imaged from the final finish.

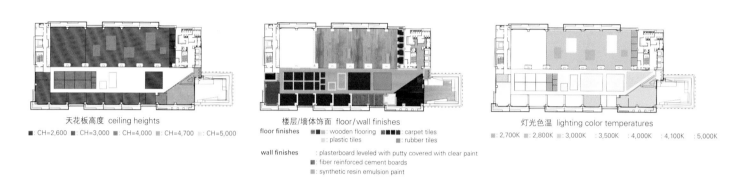

天花板高度 ceiling heights
■: CH=2,600 ■: CH=3,000 ■: CH=4,000 ■: CH=4,700 ■: CH=5,000

楼层/墙体饰面 floor/wall finishes
floor finishes ■■: wooden flooring ■■■: carpet tiles
■: plastic tiles ■: rubber tiles
wall finishes : plasterboard leveled with putty covered with clear paint
■: fiber reinforced cement boards
■: synthetic resin emulsion paint

灯光色温 lighting color temperatures
■: 2,700K ■: 2,800K : 3,000K : 3,500K : 4,000K : 4,100K : 5,000K

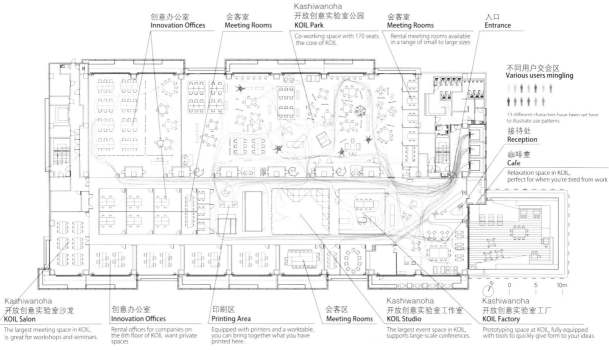

创意办公室 Innovation Offices
会客室 Meeting Rooms
Kashiwanoha 开放创意实验室公园 KOIL Park
Co-working space with 170 seats, the core of KOIL
会客室 Meeting Rooms
入口 Entrance

不同用户交会区 Various users mingling

13 different characters have been set here to illustrate use patterns

接待处 Reception

咖啡室 Cafe
Relaxation space in KOIL, perfect for when you're tired from work

Kashiwanoha 开放创意实验室沙龙 KOIL Salon
The largest meeting space in KOIL, is great for workshops and seminars.

创意办公室 Innovation Offices
Rental offices for companies on the 6th floor of KOIL want private spaces

印刷区 Printing Area
Equipped with printers and a worktable, you can bring together what you have printed here.

会客区 Meeting Rooms

Kashiwanoha 开放创意实验室工作室 KOIL Studio
The largest event space in KOIL, supports large-scale conferences.

Kashiwanoha 开放创意实验室工厂 KOIL Factory
Prototyping space at KOIL, fully equipped with tools to quickly give form to your ideas.

项目名称：Kashiwanoha Open Innovation Lab
地点：Shop and Office Tower 6F, District 148-2, Kashiwanoha Campus 178-4 Wakashiba, Kashiwa-shi, Chiba-ken, Japan
建筑师：Naruse Inokuma Architects,Co.,Ltd.
项目规划&生产：Mitsui Fudosan Co.,Ltd, Loftwork Inc.
设施设计：Kankyo Engineering INC.
照明：Izumi Okayasu Lighting Design Office
幕墙：Yoko Ando Design
家具：Battanation Co.,Ltd., Naruse Inokuma Architects,Co.,Ltd.
标志：Art direction Lemonlife&Co., Logo design MOTOMOTO inc.
施工单位：Nomura Co.,Ltd.
有效楼层面积：2,576m²
主要用途：Innovation Office
结构：SRC(steel encased reinforced concrete)
设计时间：2012.11 / 施工时间：2013.12 / 竣工时间：2014.3
摄影师：©Masao Nishikawa(courtesy of the architect)

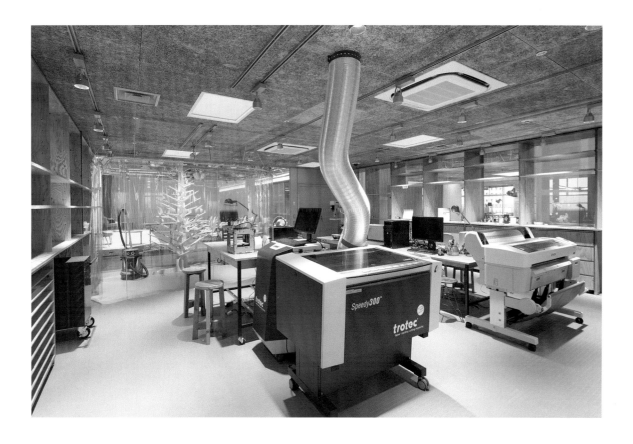

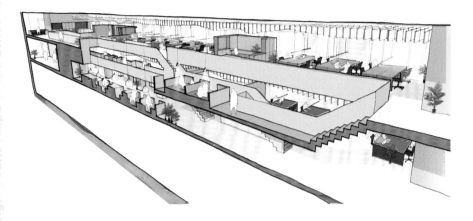

Crosswater办公室
Threefold Architects

这座桥式结构是一个新型工作环境，全长为64m，呈起伏状，多层次结构跨越两层楼的高度，位于面积为1500m²的办公和娱乐空间的双高空间内。

该桥式结构被构想为一个连续的折面，主要由预制交叉层木（CLT）建成，而且从结构方面来说，这是一个动态的外形，跨度为8m。这个优雅的元素将楼层连接起来，鼓励员工之间的互动，并且创造出一处口袋式的空间供员工们在此工作和聚会。这个桥体被设计成将四个公司统一于一个新总部的外部环境。它形成了一处鼓舞人心的工作环境。

一旦员工和客户进入一处明亮的、通风的空间，便看见了这座CLT制成的桥体结构。在这里，桥体形成了极富戏剧性的第一印象：形成休息区、楼梯和留给人印象的5m高墙，之后向上覆盖一层。

这座桥延伸至一层和二层。作为空间之间的连接元素，这座桥促成了横跨办公空间的垂直和水平方向的运动。在这座建筑中，对凸显社区的定义来说，这个联系尤为重要。从战略意义上讲，它将不同的部门和公司集合在一起。

折叠的CLT结构的设计灵感来自于历史悠久的、可居住的桥梁，经过雕琢，其上方、下方和内部都形成了空间，用于互动和聚会。这些互动的区域大小不等，从容纳1-2人的隔间到容纳40人的论坛。

折叠的木材表面被构想为桥的延伸部分，可用作柜台、存储点、隔间等。咖啡室室光线充足，有着落地玻璃窗和大阳台，是一个供人们开会、进餐、接待客人以及举办公司活动的地方。

桥的后壁以及上方的天花板形成了办公室的空间背景。建筑师给墙壁和天花板装上了48m长的、缓慢起伏的肋架。空间上方的屋顶天窗带来柔和的自然光线，光线强度通过一系列长形吊灯的作用得以加强。肋架的微波形状和光线使人联想到穿桥而过的潺潺流水，在办公环境内创造了一处梦幻般的空间。在短短的几个月内构思、开发、建造这样的一个室内设施，可谓是一个特殊的成就。它是一种新型的工作环境，一处考虑新的工作方式的灵活环境，它接纳改变，鼓励不同类型的、富有创意的商业互动。这座桥是一处优雅的动态空间，鼓励创新并激发创造力。

最重要的是，这座桥式结构给它的用户带来了幸福感，提高了员工的生产效益，同时作为公司精神富有创意的表达给客户留下了深刻的印象。

Crosswater's Office

The Bridge is a new typology of work environment, a 64m long, undulating, multi level structure spanning two floors in a double height void within 1500m² of office and recreational spaces.
Conceived as a continuous folded surface, the Bridge is constructed from pre-fabricated cross laminated timber(CLT), and is a structurally dynamic form, spanning over 8m. Bridging between floors, this elegant connecting element encourages interaction between employees and creates pockets of space in which to work and gather. Designed to be an environment which unites four companies at a new headquarters, it forms an inspirational work environment.
Both staff and guests enter into the light, airy space where the CLT Bridge structure begins. Here the Bridge creates a dramatic first impression – forming a seating area, a staircase and an impressive 5m high wall, wrapping upwards to the second floor level.

The Bridge continues to the first and second floors. As a connecting element between spaces, it encourages horizontal and vertical movement across the office. This connection is key in addressing the notion of community within the building – bringing together the different departments and companies at strategic points.

Inspired by historic inhabited bridges, the folded CLT structure is sculpted to form spaces above, below and within, for interaction and gathering. These areas for interaction vary in size from 1~2 person booths to a 40 person forum.

Conceived as an extension of the Bridge, here the folded timber surface forms counters, storage and booth areas. The cafe is a light filled space with full height glazing and a large balcony. It is a place for meeting and eating, a place to take clients and to hold company events.

The rear wall and ceiling above the bridge form the backdrop to the office space. The architects created a 48m long installation of gently undulating fins to the wall and ceiling. Roof lights over this space bring in soft natural light, which is enhanced by a series of long pendant lights. The delicate waves of the fins and light bring to mind the water that the bridge passes across, creating a dreamlike space within the office environment.

Conceived, developed and constructed in just a few months – this interior fitout is an exceptional achievement. It is a new type of work environment – a flexible environment which takes into account new ways of working – catering for change and encouraging different types of creative and commercial interaction. The Bridge is a beautiful and dynamic space which encourages innovation and inspires creativity.

Most importantly, the Bridge engenders a sense of well-being amongst its users, engineering productivity amongst employees and impressing clients as an innovative expression of the company ethos.

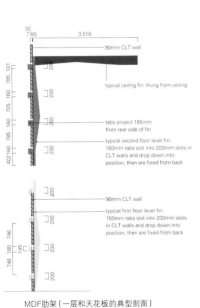

MDF肋架（一层和天花板的典型剖面）
MDF fins (typical section to first floor and ceiling)

MDF肋架的长度和测定
extent and setting out of MDF fins

- typical ceiling fin: hung from ceiling
- typical second floor level fin: 160mm tabs slot into 200mm slots in CLT walls and drop down into position
- typical first floor level fin: 160mm tabs slot into 200mm slots in CLT walls and drop down into position
- second floor level fin in front of column: column clad to front side with spruce board, fin mechanically fixed with invisible flexings
- first floor level fin in front of column: column clad to front side with spruce board, fin mechanically fixed with invisible flexings
- second floor level fin (extended length): tabs slot into 200mm slots in CLT walls and drop down into position
- short fin: biscuit jointed and glued into position
- non-typical short length first floor level fin: tabs slot into 200mm slots in CLT walls and drop down into position

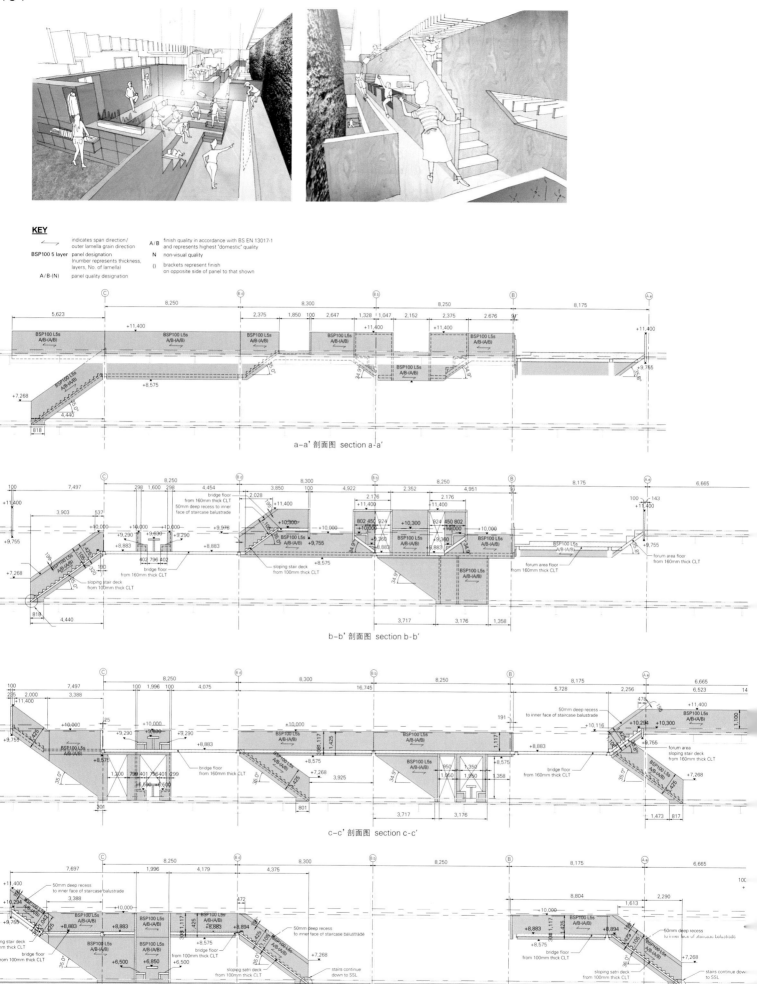

项目名称：Crosswater Company
地点：Lake View House, Rennie Drive, Dartford, Kent, DA1 5FU, UK
建筑师：Threefold Architects
主要开发商：Prologis
项目经理：Savills
结构工程师：
RPS Planning & Development _ main building,
Aecon _ bridge
主要承包商：Winvic
CLT承包商：Construction al Timber
建筑服务：Walter Miles Electrical Engineers Ltd
建筑控制：Salus
有效楼层面积：entrance _ 90m² , 1st & 2nd floor _ 1,400m² ,
cafe _ 200m² / 桥体长度：64m
甲方：Crosswater
竣工时间：2014.7
摄影师：©Charles Hosea Photography(courtesy of the architect)

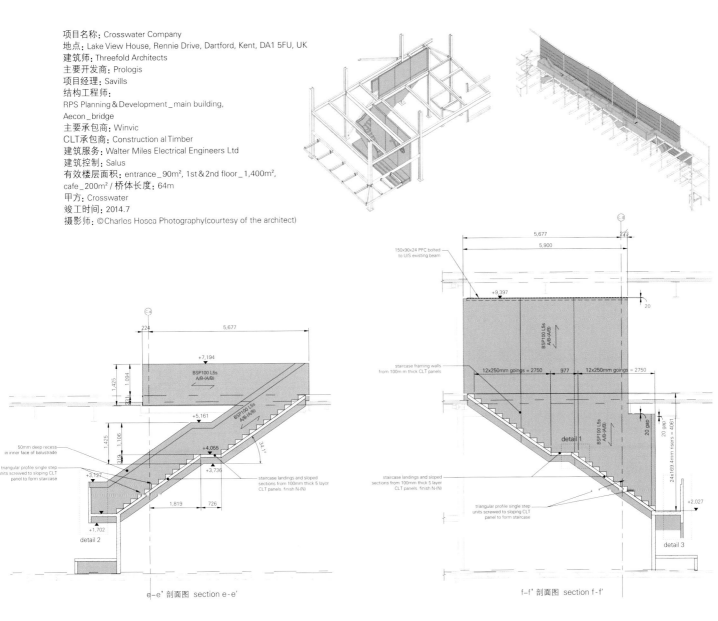

e-e' 剖面图 section e-e'

f-f' 剖面图 section f-f'

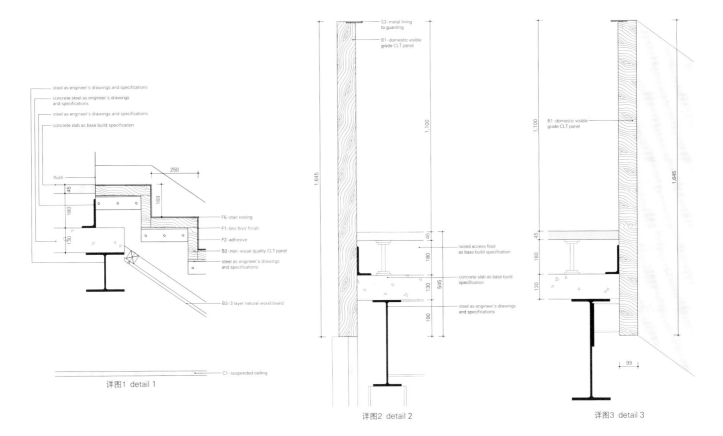

详图1 detail 1

详图2 detail 2

详图3 detail 3

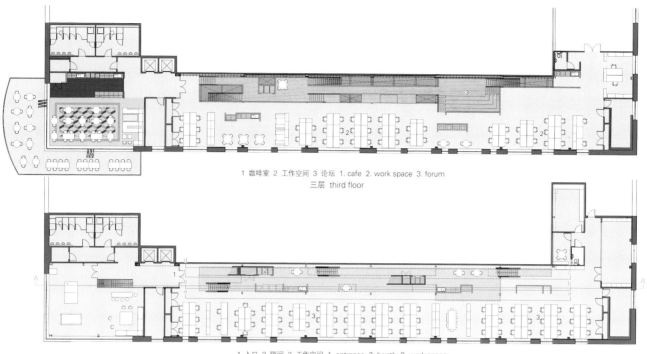

1 咖啡室 2 工作空间 3 论坛 1. cafe 2. work space 3. forum
三层 third floor

1 入口 2 隔间 3 工作空间 1. entrance 2. booth 3. work space
二层 second floor

一层 first floor

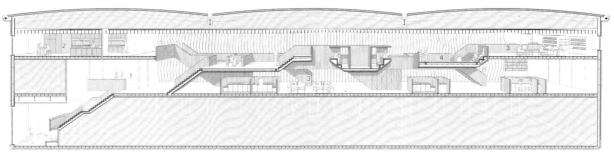

1 入口 2 咖啡室 3 工作空间 4 论坛 1. entrance 2. cafe 3. work space 4. forum
A-A' 剖面图 section A-A'

马德里马塔德罗文化工厂
Office for Strategic Spaces

在遭受危机重创的欧洲,尤其是西班牙,设计师们不得不寻求出路,来解决资金匮乏与巨大需求之间的对立冲突。本案或许是其中的典型代表。建筑师使用了极少的、价格低廉且易于安装的材料,并且设法利用这些材料打造尽可能多的不同办公领域,以满足不同的需求。

靠近入口处的三个建筑体量不但形成了空间布局,而且对周围的步道进行了折叠和压缩。由此构建从紧凑到开阔、从繁华到静寂的过渡区,以实现工作空间的多样性。在历时不到一个月的时间里,建筑师以每平方米105欧元的价格,完成了这项修复性的、空间层叠的项目,并且使其适应甲方所要求的众多工作场景。

如果只利用平层区域,在399m^2的空间内划分出所需的120处办公空间是不可能实现的(若除去公共通道占地,可使用的实际建筑面积需要进一步减少到340m^2),除非我们能够利用建筑的挑高,来获得更多的可用空间。然而甲方并没有足够的经费来建造传统意义上的二层,新的问题由此产生。于是建筑师决定采用最为简单的一种建造体系:使用最廉价的本地松木,全部采用相同的标准尺寸,以简化建筑结构的供给和施工流程;选用重量轻、尺寸大的多层聚碳酸酯板,这样,墙面的施工仅用一天就完成了。这样不仅多出了85m^2的面积,至关重要的是,建筑师还分割出两层的功能空间,因此在空间使用上更为灵活,现在甲方对这些空间也进行了很好的利用。

建筑师建造了三个小型体量,或者说三座小型建筑物,它们改变且改善了可用空间的品质,并且相应地构建了建筑师认为十分必要的多样工作区域。这三个风格鲜明的建筑体量靠近项目的入口处,有助于对功能进行规划,且折叠和压缩周围的步道,使它们尽可能的精准而紧凑。这样的布局不仅通过减少流线面积来节约空间,而且还实现了从紧凑到开阔、从繁华到静寂的过渡,有助于实现工作环境的多样性。

建筑方法的缺乏导致建筑师只能在空间直觉上进行进一步拓展:似乎在那些未设计完成的、未完工的环境中和处于设计进程之中的空间最能够激发创造性工作和工作者的灵感。建筑方法的缺乏还让建筑师得以在建筑中进行一些颇有示范意义的小实验:包括在楼板和扶手之间的垂直木构件里安装灯具;设置极其简单的吸音系统来改善建筑内部的声环境,而这原本要依赖厚重的、传统的、昂贵的墙体构造才能实现。

建筑师主要担心的是如何把项目设计变为可能,并且使设计成果既经济又可持续。建筑师很高兴最终以每平方米105欧元的造价实现了建筑基本的可持续功能——以非常低廉的建造费用产生了较高的社会影响及实用效果。

项目采用的建造策略非常明智,选用当地最经济的松木,使用具有可持续性的森林木材。建筑结构总共吸收将近880吨二氧化碳。

多花一点时间精算数据、多花一点精力诠释空间的品质(设计成开放的空间、挑高的天花板和整齐的通道),这样建筑师就有能力在保证建筑安全的情况下,尝试不给木材喷刷耐火的化学材料涂层,使建筑环境免受有毒化学物质的影响。

Cultural Factory in Matadero Madrid

The project is perhaps typical case in crisis-stricken Europe, and particularly Spain, where the architect has to find a way out of the opposing tensions of having no money and big needs. The architects used very few, cheap, and easy-to-install materials, and they tried to achieve with them as many different and distinct work areas as possible, adapted to different needs.

Three volumes near the entrance organize the space, folding and compressing the circulations around it. This creates a gradient, from compact to expansive, from busy to silent, that helps achieve variety in work spaces. In a little less than one month the architects built a reversible, vacuum-packed, EUR 105/m² project, adaptable to the multitude of situations the client asked for.

To house the needed 120 work spaces in a 399m² area, a floor area (that needed to be further reduced to 340m² in order to maintain a public pass-through) was impossible, unless the architects found more spaces using the height of the hall. This created additional problems since there was not enough money to achieve the construction of a second floor by traditional means. The architects decided to use very simple building systems: the cheapest local pine lumber, all in the same standard size, which simplified the supply and construction process of the structure, and multi-wall polycarbonate panels, very lightweight and in large sheet, which allowed for the walls to be finished in just one day. They were able to achieve 85m², and crucially to split functions in two levels, which allow for more flexibility in use that the client is now making very good use of.

They could build three small volumes, or architectural objects, that changed and modified the quality of the available space, naturally producing the diverse working areas they thought

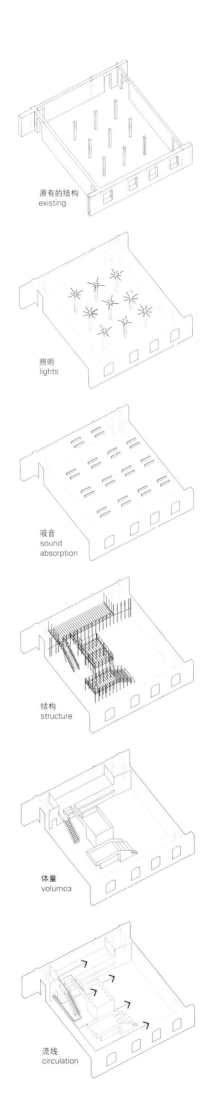
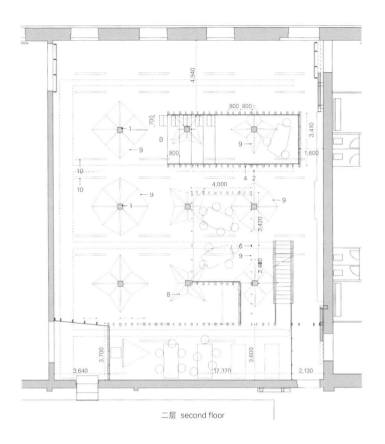

二层 second floor

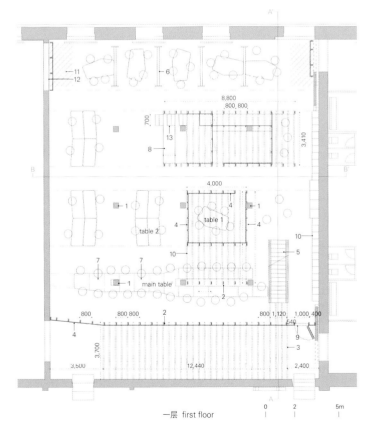

一层 first floor

1. existing structure 2. uprights, sawn and brushed down pine wood, 6x18cms. approx. 3. projected pine wood beams, sawn and brushed down, 6x18cms. approx. 4. cellular polycarbonate 32mm, bolted to wooden structure 5. stairs to loft, sawn and brushed down wooden elements, post-tensioned to base of bolts, according to plan D3 6. mobile spacers, according to plan 7. common workbench according to plan D2 8. bleachers according to plans, the same structure as the general element 9. access door, pine wood frame, cellular polycarbonate, opening and closing system to be defined in conjunction with related access control systems 10. lockers 11. areas reserved for emergency evacuation 12. door for emergency evacuation, opening system coordinated with general alarm 13. plywood box steps

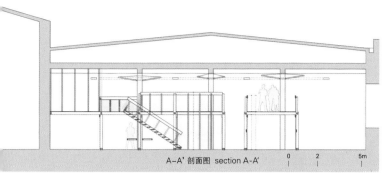

A-A' 剖面图 section A-A'

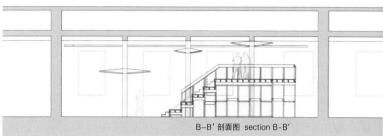

B-B' 剖面图 section B-B'

necessary. These three distinct volumes near the entrance help organize the program, folding and compressing the circulations around it, making them as exact and compact as possible. This organization not only saves space by reducing circulation areas, but creates a gradient, from compact to expansive, from busy to silent, that helps achieve the needed variety in work environment. The scarcity of means allowed the architects to research one spatial intuition: creative work, and workers, thrive in environments that seem not completely designed, not completely finished, and spaces that seem to be caught in the middle of a process. It also allowed for small experiments in architecture, with demonstrative qualities. Among them: the light fixtures are made to the vertical structural wood elements between the floor and the handrail; a ridiculously simple sound absorbing system corrects the sound conditions that would otherwise rely on heavy, traditional, and unaffordable wall construction.

The architects' main worry was to make it possible, and to achieve that they needed to make it financially responsible and sustainable. At EUR 105/m². The architects are very happy to have helped its basic architectural sustainability: to achieve a high social and usable impact for the money.

In the wise construction process, the most economic pinewood from local, sustainable forests was used. Nearly 880 tons of CO_2 are captured in its structure.

A little more effort at the time of calculation and the ability to explain its spatial qualities (the fact that the space is maintained open and high-ceilinged, and circulations clear) allowed the architects to argue the possibility of keeping the wood untreated with fire-retardant chemicals, which makes the environment free of noxious chemicals. Office for Strategic Spaces

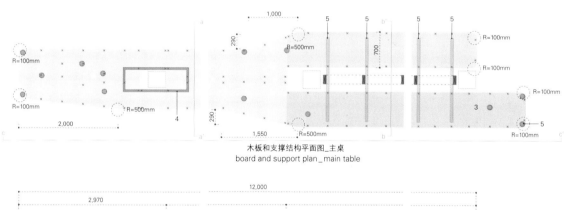

木板和支撑结构平面图_主桌
board and support plan_main table

1. plywood 15 mm
2. pine wood frame 60x35 mm
3. plywood panel 244x122mm
4. triangular support, laminate steel rod 40,40,4
5. support beam made of sawn and brushed down pine wood, squaring 180x60, lowered according to plan, bolted to the pillar with steel plate and screws alike detail of structure (x6)
6. varied table legs according to plan (source: Ikea)
7. existing pillar

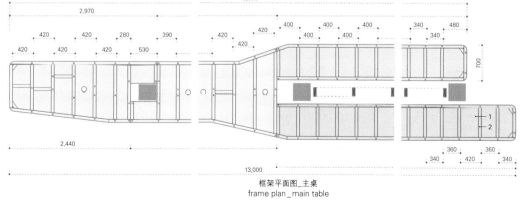

框架平面图_主桌
frame plan_main table

a-a' 剖面图 section a-a'

b-b' 剖面图 section b-b'

c-c' 剖面图 section c-c'

项目名称：Cultural Factory Matadero Madrid. Creative Hub
地点：Paseo de la Chopera 14, Madrid, Spain
建筑师：Office for Strategic Spaces
首席建筑师：Angel Borrego Cubero
结构工程师：MOZ
建造工程师：Dulsberg SL
用地面积：484m²
总建筑面积：399m²
有效楼层面积：484m²
造价：EUR105/m²
设计时间：2013 / 竣工时间：2014
摄影师：©Simona Rota(courtesy of the architect)

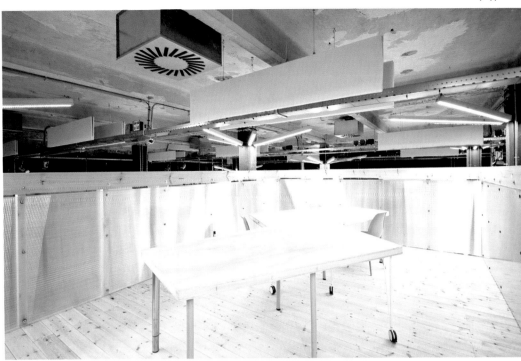

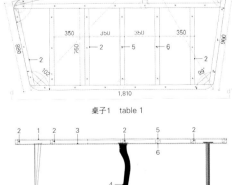

桌子1　table 1 桌子2　table 2

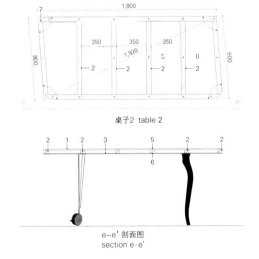

d-d' 剖面图
section d-d'

e-e' 剖面图
section e-e'

1. plywood 15 mm
2. pine wood frame 45x35mm
3. plywood 15 mm
4. varied table legs according to plan (source: Ikea)
5. brass flat head screw 3.20
6. brass flat head screw 3.15
7. rounding radius 10cms
8. rounding radius 30cms

IBC创意工厂
Schmidt Hammer Lassen Architects

IBC创意工厂的设计支持全新的学习方式。面积为12 800m²的教学大楼是由一座1978年建造的GORI油漆厂翻新而来,在当时,这座工厂为其他工厂的建造建立了新标准。Schmidt Hammer Lassen建筑师事务所秉承老工厂的精神,与位于柯灵的国际商业学院(IBC)合作,为创意的学习环境营造背景,旨在成为世界首屈一指的教学建筑。该建筑的设计目标是成为未来创意者的培训大本营。

在2010年夏天得到GORI工厂的所有权之后,国际商学院打开了通往富有开拓精神和视野的物理环境的大门。这是丹麦第一家将生产与管理集合在一个大房间的工厂,并且二者之间具有视觉连接。大型油漆罐由法国艺术家Jean Dewasne进行了装饰,旨在使这处工作区内的艺术能

项目名称:IBC Innovation Factory
地点:Kolding, Denmark
建筑师:Schmidt Hammer Lassen Architects
工程师:Rambøll A/S
消防顾问:Alectia
主要承包商:MT Højgaard a/s, Odense
甲方:IBC International Business College, Kolding
总建筑面积:15,200m²
施工时间:2010—2012
摄影师:©Adam Mørk(courtesy of the architect)

够激发员工的灵感,以提供更好的工作环境。

在生产楼层,为员工设置的羽毛球场和乒乓球桌也是由同样的理念激发的。Schmidt Hammer Lassen建筑师事务所的主要任务是保护和强调建筑原来的质量,并将原有的设施转化为创意学习中心。

通过使用六个元素,即火、水、植被、光、声音以及空气,全新的创意工厂理念的开发集中在用户的感官方面。这座中心教学设施的外形如同室外的"景观家具",材质为花旗松,将各样的学习空间和学习体验融合在一起。木结构看起来是漂浮在水面上的,内设礼堂、开放的学习区、圆形广场,以及封闭的、用于集体工作和自习的矮墙。此外,这里还有植被、滴水、飞鸟(在鸟舍中),以及从顶部天窗射来的光线。

IBC Innovation Factory

The IBC Innovation Factory is designed to support new ways of learning. The 12,800m² educational building is the result of a refurbishment project of the paint manufacturer GORI's factory from 1978, which set new standards for factories at the time. In the spirit of the original factory, Schmidt Hammer Lassen Architects, in collaboration with International Business College(IBC) Kolding, has created the settings for a creative learning environment, aiming to become the world's best. The ambition is to be a training

camp for future innovators.

With the acquisition of the GORI factory in the summer of 2010, the IBC gained access to physical environment characterized by pioneering spirit and vision. It was the first factory plant in Denmark to unite production and management in one large room, allowing visual connection between the two. The large paint tanks were decorated by the French artist Jean Dewasne, in the conviction that art in the workplace would inspire employees and provide a better working environment. The same idea inspired the incorporation of badminton courts and Ping-Pong tables on the production floor for the employees.

The main task for Schmidt Hammer Lassen Architects has been to preserve and emphasize the building's existing qualities and transform the facilities into an innovative learning environment. By using six elements – fire, water, greenery, light, sound and air – the concept for the new Innovation Factory was developed with an emphasis on stimulating the users' senses. A central teaching facility in the shape of an indoor "landscape furniture" designed in Douglas pine wood incorporates a variety of learning spaces and experiences. The wooden structure, which seems to float above a surface of water, has an auditorium, places for open study, an amphitheater, and closed podiums for group work or quiet study time. There are green plants, the sound of trickling water, birds in aviaries and plenty of daylight from the skylights above.

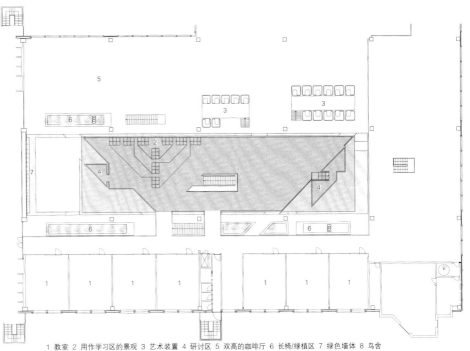

1 教室 2 用作学习区的景观 3 艺术装置 4 研讨区 5 双高的咖啡厅 6 长椅/绿植区 7 绿色墙体 8 鸟舍
1. classroom 2. learning landscape 3. art installation 4. workshop areas
5. double high cafe area 6. bench/green plants 7. green wall 8. aviary
二层 second floor

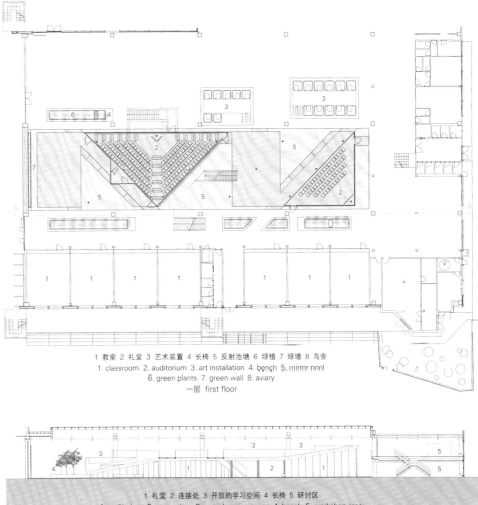

1 教室 2 礼堂 3 艺术装置 4 长椅 5 反射池塘 6 绿植 7 绿墙 8 鸟舍
1. classroom 2. auditorium 3. art installation 4. bench 5. mirror pool
6. green plants 7. green wall 8. aviary
一层 first floor

1 礼堂 2 连接处 3 开放的学习空间 4 长椅 5 研讨区
1. auditorium 2. connection 3. open learning space 4. bench 5. workshop area
A-A' 剖面图 section A-A'

Lowe Campbell Ewald总部

Neumann/Smith Architecture

Lowe Campbell Ewald广告公司为一座有着100年历史的老建筑注入了新的活力，开创了长期闲置的底特律的老建筑再利用的先例。建筑改造既保护了历史建筑文物，也使创意产业区成为底特律发展的强大经济引擎。

2012年，Lowe Campbell Ewald广告公司为总部迁址考察了若干套设计方案，最终选定这处毗邻密西根州底特律市中心福特球场的J.L.Hudson公司旧仓库。

J.L.Hudson公司仓库修建于20世纪20年代。它是多种结构体系的一个结合体（随着时间的推移，在老建筑的基础上增建了若干座建筑），既使用了粘土砖瓦，也有成形的钢筋混凝土平板屋顶。现有的所有建筑元素在Lowe Campbell Ewald公司新的设计方案中都显露出来。施工范围包括五层通高的开放办公区，新的机电系统，以及一间配有专用不间断电源设备、空调系统和备用发电机的先进的计算机机房。按照以前签署的合同，Lowe Campbell Ewald公司大楼拥有一个五层高的巨大中庭，贯穿建筑上部的四层楼面，这个中庭有别于底特律的任何其他建筑，因此成为项目的一个重要结构特征。

Lowe Campbell Ewald公司选择配置长凳家具，并且尽可能地减少固定房间的数量，以打造更加开放、更富有协作性的办公环境。建筑设计

富有创意地使用回收的旧建筑材料,包括木质工作台、在当地搜集的500扇旧木门(切割成隔间的板条)、以及用作空间隔断的电气管道。空间的特色还在于采用一些可持续的建筑设计方案,包括回收利用来自密西根谷仓的木材、混凝土制作的吧台、再生材料和环保面料制成的椅子。

办公室的饰面设计也独具特色。门厅天花板的装饰采用可以追溯到数码印刷产品出现之前的黄铜印刷板,每块板面上都印有Lowe Campbell Ewald公司20世纪50年代到80年代的广告,利用新办公空间来增强其个体形象。

一个四层楼高的LED屏幕上能够播放定制的广告词和图像,提供视觉上的冲击效果。Neumann/Smith建筑事务所设计该屏幕的目的是为了填充中庭的高度,同时制造出令人发出惊叹的"哇"的效果。12.8m高的显示屏主要供三楼中庭的访客和员工观看,这里专门为个人或者公司聚会设置了不同的平台。另外,公司的品牌渗透区包括一系列的多触感互动屏、LED照明设施、新浴室、小厨房和精心设计的室内球场也极大地提升了环境氛围。

电路的设计和施工也是建筑的重要构成要素。建筑内部的一些视觉冲击区域使用再生荧光灯管打造而成的独特的照明装置,而开放的办公区域则主要选用具有时尚、节能品质的工作照明灯。

Lowe Campbell Ewald Headquarters

Advertising agency, Lowe Campbell Ewald, has breathed a new life into a 100-year old building, setting a precedent for repurposing long vacant Detroit buildings, shining the light on historic preservation and anchoring the creative sector's place as a strong economic engine for Detroit.

In 2012, advertising agency Lowe Campbell Ewald examined several options for relocating their corporate headquarters before falling in love with the "bones" of the former J.L. Hudson Co. warehouse connected to Ford Field in downtown Detroit, Michigan. The J.L. Hudson Co. warehouse building dates back to the 1920s. It features a combination of multiple structural systems(due to several additions over time) with clay tile and board-formed concrete and steel deck ceilings – all which were called out to be exposed in Lowe Campbell Ewald's new design. The scope of work included five full floors of open office space, new mechanical and elec-

trical systems and a sophisticated computer room with dedicated UPS, HVAC and backup generator. A five-story atrium, unlike any in Detroit, was carved from the upper four floors under a previous contract and serves as a key architectural feature.

Bench furniture configurations and minimal fixed rooms were selected by Lowe Campbell Ewald to create a more open and collaborative office environment. The design also makes creative use of recycled materials including wooden pallets, 500 locally salvaged wood doors sliced up into slatted partition pods, and electrical conduits used as room dividers. The space features other sustainable solutions including reclaimed barn wood from Michigan, counters made of concrete, and chairs with recycled content and environmentally friendly fabrics.

The office features highly unique finishes. Brass press plates dating back to an age before digital print production adorn the lobby's ceiling. Each plate features a vintage Lowe Campbell Ewald advertisement from the 1950s to the 1980s and strengthens their personal identity with their new space.

A four-story LED screen with the ability to add customized messaging and imagery provides visual impact. Neumann/Smith Architecture designed the screen to fill the height of the atrium and created a "WOW" factor. The 42-foot high display screen engages visitors and employees from the third floor atrium, which is designed as tiered platforms for personal and company gatherings. Additionally, the company's brand immersion area includes a series of multi-touch interactive screens. LED lighting, new bathrooms, kitchenettes, and highly designed "pitch rooms" enhance the environment.

Electrical design and construction were also significant components. Unique lighting fixtures were created in high impact areas that incorporate recycled fluorescent tubing. Task lighting featured in open office areas was selected for its sleekness and energy saving qualities.

项目名称：Lowe Campbell Ewald Headquarters
地点：Detroit, Michigan, U.S.A.
建筑师：Neumann/Smith Architecture
平均楼层面积：2266.83m²
有效楼层面积：11,334.17m²
总建筑规模：five stories above ground
造价：USD 15,000,000
竣工时间：2014.1
摄影师：©Justin Maconochie(courtesy of the architect)

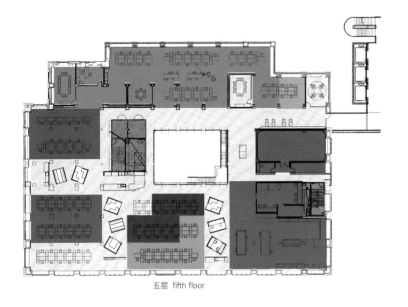

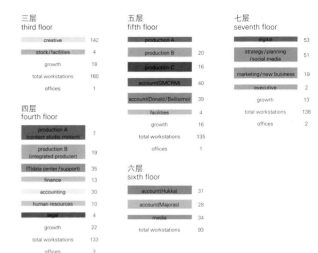

三层 third floor		五层 fifth floor		七层 seventh floor	
creative	142	production A		digital	53
stock/facilities	4	production B	20	strategy/planning /social media	51
growth	18	production C	16	marketing/new business	19
total workstations	160	account(GMCRM)	40	executive	
offices	1	account(Donald/Bellisimo)	39	growth	13
四层 fourth floor		facilities	4	total workstations	138
production A (contect studio motion)	7	growth	16	offices	2
production B (integrated producer)	19	total workstations	135		
IT(data center/support)	35	offices	1		
finance	13	六层 sixth floor			
accounting	30	account(Hukka)	31		
human resources	10	account(Majoras)	28		
legal	4	meda	34		
growth	22	total workstations	93		
total workstations	133				
offices	3				

五层 fifth floor

四层 fourth floor

七层 seventh floor

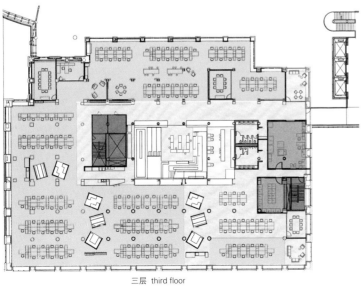

三层 third floor

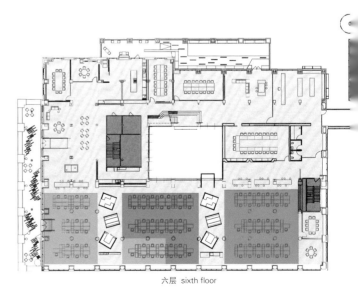

六层 sixth floor

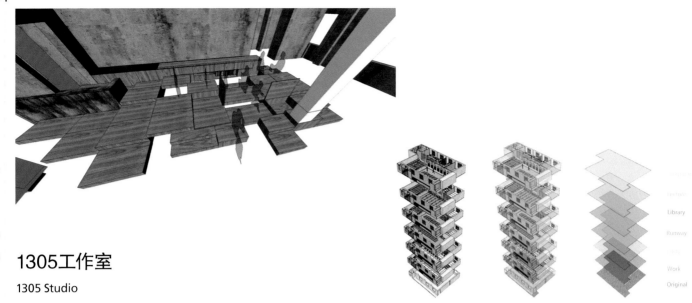

1305工作室
1305 Studio

项目名称：1305 Studio Office
地点：Room 210, No. 383, Changhua Rd, Shanghai, China
建筑师：1305 Studio
用地面积：306m² / 总建筑面积：1,500m² / 有效楼层面积：1,200m²
设计时间：2013 / 竣工时间：2014
摄影师：©SHEN Photography (courtesy of the architect)

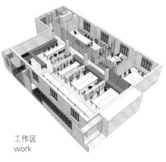

工作区
work

讲座区
lecture

图书馆
library

　　位于上海市静安区石库门的1305工作室被设计成充满创意的多功能空间。这里不仅可供建筑设计、室内设计和平面设计使用，还可以举办时装秀、艺术展、鸡尾酒会、专业讲座等活动。建筑师对上海的弄堂文化有着深厚的兴趣，因此致力将这个306m²的空间打造成融传统元素与现代气息为一体的和谐空间。结果证明，这的确是个不错的尝试。

　　空间的关键词是盒子。自由堆积或散放的盒子随处可见，为该空间提供了无限的可能性。在这里，人们可以轻松地适应环境，找到恰当的方式与人交流，分享快乐。

　　设计的灵感来源于玻璃杯的启发。工作室的临时使用空间正如一个不受所谓"空间界限"限制的玻璃杯，充满无限可能。如同"一杯水"，如果倒掉杯中水，反而开启了无限的可能性，把不同的物质倒入杯中，就会得到新的名称，也许是"一杯牛奶、一杯果汁、一杯啤酒"或者一杯其他的东西。

　　工作室的空间如同一杯水，工作时，我们便在这个"杯"中看到所谓的办公桌、文件柜等办公用品。但无论是办公桌还是文件柜，设计师都会设计一个精算好的空间模数，使其能够自由组合，满足未来五年甚至十年后的办公空间的使用需求。

　　每到周末，办公空间会彻底变换成另一番景象：办公用的前台接待台变成酷炫的DJ台，或者摆满各种酒水饮料的吧台，7m长的投影屏演绎各种光怪陆离的影像。

　　设计师们用木盒搭建走秀的T台，周末可以邀约好友聚会，举行小型的时尚秀，自娱自乐。

　　每到周末，平日的办公室便不再是工作的空间，而成为一个展厅或艺术展览会。

　　堆叠起所有的书架，室内空间就会变身成为一个小型图书馆。两种不同模数的白色木架能够组合成观众所需的坐凳，满足50～100人左右

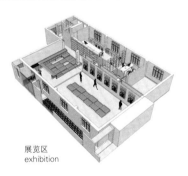
展览区
exhibition

跑道
runway

聚会区
party

的小规模演讲报告会使用。

　　橡木质地的木盒具有存储功能。通过反复演算，设计师们设计出这些尺寸精准的盒子，精确的尺寸确保建筑空间能够进行不同的功能展示。盒子的高度分为15cm和30cm两种，它们可以自由组合和叠加，叠加的高度为45cm、60cm、75cm或90cm等以此类推的更高尺寸。木盒门开启方向的设计兼顾组合之后空间变化的差异性与隐私性。书架可以拆分成宽度为15cm和30cm的两种不同的单体，两者错落叠加可以增强书架的稳定性。传统的储物间、书架均可拆解为小单元结构，充分考虑到流线、环保和可持续发展等因素。

1305 Studio Office

Located in Shikumen building in Jingan District, Shanghai, 1305 Studio Office is set to be the space with multi creative functions. The space acts not only for architecture, interior or graphic design, but also for lots of other purposes such as fashion show, art exhibition, cocktail lounges, professional lectures, etc. With the interest of Shanghai Lane Culture, the architects tried to make this 306m² space into a harmonious combination of tradition and modernism. The result turns out that it was not a bad attempt.

The language of the space is the box. Boxes lying around, either piled up or separated freely, make the space of unlimited possibilities. People in the space can easily find a way to adapt and communicate properly, and share joy with others.

The concept of this design was inspired by a glass cup. The temporary living space is just as a glass cup without the so-called

"space border", but with many limitless possibilities. Just as "a glass of water", when it falls into the water, on the contrary it will get limitless possibilities. Putting different substances into the cup will have a new name, may be "a glass of milk, juice, beer" or others.

Work space is just like a cup of water, when we work in this "cup", we can see the so-called desks, file cabinets, etc. But for both the desk and file cabinets, designers designed space accurately, so that it could be free to combine and to meet the demands over the next five years or even ten years later.

Every weekend, the office space becomes an entirely different place. It may use the front desk into a cool DJ counter, or a counter with all kinds of drinks, and a 7 meter-long screen shows various psychedelic images.

Designers design T station with wooden boxes. At weekends, people can get together with friends holding a small fashion show.

At weekends, when the office space is no longer a working space, it turns into an exhibition hall or an art exhibition.

When all the bookshelves are folded, it will turns into a small library. Two white wooden supports of different modulus can turn into a speech sitting stool. It could meet a small speech with about 50-100 people.

The wooden box, which is made of oak, has storage capabilities. Through repeated calculations, designers can design the boxes of accurate size. The accurate size enables the space to show different functions. The heights of them are 15cm and 30cm. They can freely combine and superpose. They could be superimposed to 45cm, 60cm, 75cm or 90cm and so on. The open direction of the wooden door both considers spatial variability, and also the privacy. The shelves could be divided into units with a width of 15cm and 30cm, staggered and superposed, to increase the stability, in the meanwhile, in consideration of the circulations, environment protection, and sustainable development, the traditional lockers and bookshelves could be divided into small units.

工作区 work　　　　　　讲座区 lecture　　　　　　图书馆 library

展览区 exhibition　　　　跑道 runway　　　　　　聚会区 party

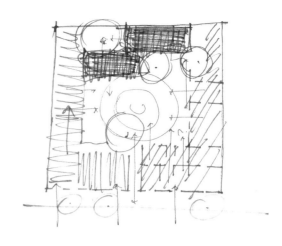

CDLE办公室

R-Zero Arquitectos

该办公室的主建筑建于20世纪初,鉴于其艺术价值、历史价值以及民族价值,被INBA、INAH和SEDUVI (均为政府机构) 列为历史文化遗产收录在册。建筑最初作为住宅使用,后来又被用作面包房、妓院等各式商业用途,历尽铅华。这些经历使建筑的本质得以升华,赋予其独一无二的魅力。建筑如同一个真实的人,所有的经历磨砺出它的个性与性格,以其特有的伤疤、标志和印记造就一处非凡的景致。这些特有的印记为人们所敬重,也成为建筑的重要组成部分,陪伴它开启生命中新的篇章,投入到不同的用途之中。

多年来,项目的选址几经周折和变迁。20世纪初,项目所在的地区主要是休闲客栈或者有钱人的小别墅,但是由于城市的发展并不理想,地区所处地理位置也并非市中心,因此这个地区慢慢被大家所遗弃。1985年,墨西哥城大地震引发了不小的灾难,这也对城市的动态分布产生了消极的影响,使得这片充满历史积淀的区域再难以融入都市生活的大环境中。

由于已被INBA所收录,该建筑的建造过程必须保留原有建筑的本质风格,同时,整栋建筑要有灵活的使用空间和宽广的全景视野,以凸显主入口广场的重要性,使每位客户发现其投资价值,进而满足三方不同客户的需求。

从设计第一份建筑草图时起,设计师一直在尝试建造中空的建筑体块;这些建筑体块的排列分布和高度都根据中心广场来制定。按照这种方法,我们最终建成了亦静亦动的空间。

建筑内部的每个区域都有许多不同的结构,这些结构并非人为建造,而是随着时间的推移逐渐产生并显现出来的。废旧的颜料、裸露的砖块、攀越老旧的房梁和墙壁的古老植物,不管这些过去曾经记录下什么,现在都被当作时光的雕刻品。建筑师不对这些做任何改变,设计的唯一限制就是建筑本身的保存问题。

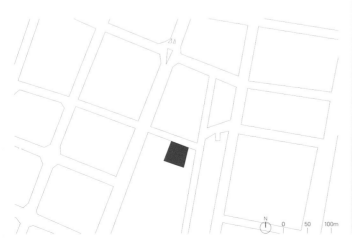

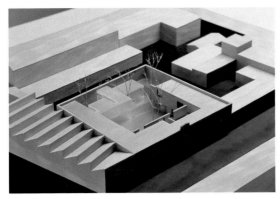

花园的美景让人沉醉，素净的颜色和质地与周围裸露的红砖恰好形成对比。花园中的三处文化遗址见证了建筑的历史，凸显出该建筑真实的品质。

在这个砾石花园中，新遗址很突出，且彼此相隔而立，在建筑内部的每一个重要景点上都可以看到它们。石质雕塑以其不规则的、坚固的形象为建筑增色不少，老旧木梁改建而成的长凳为建筑内的人们提供冥思、小憩之处。

CDLE Office

The main building was constructed in the beginning of the 20th century and it is cataloged by INBA INAH and SEDUVI (government organizations) as a historic property because of its artistic, historic and national value. It was destined first for housing use but later in time it hosted different businesses from a bakery to a brothel. All the experiences are what give the building its essence and make the space unique. The building acquires its personality and character as a real person, with all those experiences, in a practical way which results in a singular place with its own scars, marks, and prints, which will be respected and will take an important part in the project, and accompany the new chapter of its life, into different uses.

The site of the project has encountered and suffered many changes over the years. In the beginning of the 20th century, the district

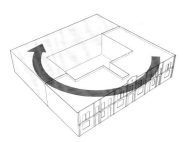 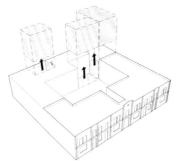 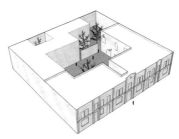 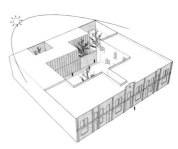

对原有的体量进行完善
原有的区域全部反转,以利用露台体量的优势,使建筑结构变得紧凑,并始终寻求与中央天井相联系。
completing the pre-existent volume
the total pre-existent area is inverted to take advantage of the terrain capacity, leaving the building structure intact and always looking for a relation with the central patio.

体量规划
将私人花园向上移,形成其下容纳不同空间的规划。
volumetric proposal
extracting the blocks of the private gardens, creates a proposal that lets the different spaces breath.

与室外连接
项目寻求与室内花园和主广场之间的联系。根据功能需要,利用不同的路线,空间可以获得更多的隐私。
connections with the exterior
the project seeks a visual relation to the internal gardens and the main plaza. with the different routes, the spaces gain more privacy than others if the program demands it.

最终的规划
最终的规划是设有一个外立面,立面使中央天井扩大一倍。原始的建筑利用原始墙体上的几个嵌入结构而得以保留。
final proposal
the final proposal has an exterior facade that duplicates the central patio. the original building is conserved with only a few interventions on the original walls.

was mainly filled with resting houses or cottages with wealthy owners, but the city's growth and its location near the heart of the city caused the abandonment of the area. The disaster provoked by Mexico City's earthquake on 1985 triggered a negative effect in the city's distribution dynamics and made it difficult for the historical districts to consolidate in the urban context.

Because of being cataloged by the INBA, the intervention in the construction must try to keep the essence of the preexistent building. It responds to the requirements of three different clients, flexible spaces and a wide visual range of all the area, which wants to give each client a feeling of appropriation of space giving a special importance to the main entrance plaza.

Since the first sketches, the architects were searching to generate solid blocks with voids; the arrangement of these blocks and its height are based on the central plaza. With this actions we have static and dynamic spaces as a result.

The building interior has plenty of different textures in every area. Textures cannot be produced artificially, they are only what time can generate and will not be hidden. It doesn't matter if its wasted paint, raw brick, old vegetation that has overcome some old beams or walls, used to carry something before, nowadays they work as almost sculptures, nothing is changed, the only limitation is the preservation of the building itself.

The garden is mainly contemplative, and its sober colors and textures contrast directly with the raw red of the bricks that surround it. Three ashes that are part of the building's history reveal its true character.

Young ashes are the new accents of the gravel garden that rise apart from each other, and can be seen from strategic points of the building interior. The stone monoliths enrich the space with its irregular and solid figures. Old wooden beams that were recuperated work as benches help contemplate and rest inside the space.

一层 first floor

二层 second floor

项目名称：CDLE Office
地点：Mexico City , Mexico
建筑师：Alejandro Zarate de la Torre, Edgar Velasco Casillas
项目团队：Mario Pliego, Eliud Martinez, Norma Contreras, Didier Lopez
施工单位：Colectivo A Heriberto Maldonado, Alfonso Baez
景观建筑师：PAAR Carlos Alberto & Paola Lopez
家具设计：exterior _ Ariel Rojo, interior_Perigonal
用地面积：1,894.69m² / 施工面积：2,330.43m²
设计时间：2011 / 竣工时间：2014
摄影师：
©Moritz Bernoully(courtesy of the architect) - p.172~173[bottom], p.174, p.175, p.177, p.178[top], p.179[top]
©Jaime Navarro(courtesy of the architect) - p.173[top-left], p.176, p.178[bottom], p.179[bottom]

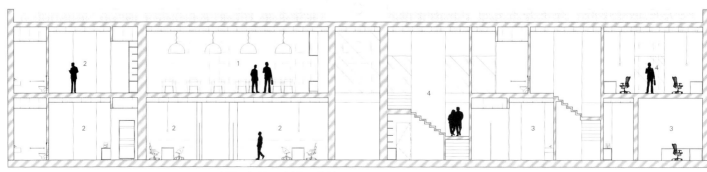

1 公共区 2 办公室1 3 办公室2 4 办公室3   1. common area 2. office 1 3. office 2 4. office 3
A-A' 剖面图 section A-A'

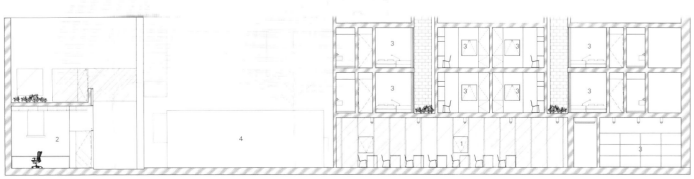

1 公共区 2 办公室1 3 办公室2 4 天井 1. common area 2. office 1 3. office 2 4. patio
B-B' 剖面图 section B-B'

>>70
architectuurstudio HH
Herman Hertzberger[left] founded his office in 1958, Amsterdam. Since 2007, the office has been operating as architectuurstudio HH by Herman Hertzberger, Patrick Fransen[right], Laurens Jan ten Kate. They aim to give every building a distinctive character which emphasizes the client's identity and try to make a significant contribution to the sustainability of our environment. So they distinguish three major aspects of sustainable building design; a long physical lifetime, an energy efficient installation concept and a culturally sustainable for functional change.

>>40
Coop Himmelb(l)au
Was founded in 1968 by Wolf D. Prix, Helmut Swiczincsky and Michael holzer in Vienna, Austria. Is active in architecture, urban planning, design, and art. In 1988, a second studio was opened in LA, USA and further project offices are located in Frankfurt, Germany and Paris, France. Recently, he won the MIPIM Architectural Review Future Projects Award in the sustainability category for the Town Erdberg and received the Wallpaper Design Award 2011 in the Best Building Sites category.

>>58
Studio Libeskind
Daniel Libeskind was born in Poland, 1946 and became American citizen in 1965. Received a postgraduate degree in History and Theory of Architecture at Essex University in England, 1972. Has designed world renowned projects including the master plan for the World Trade Center in New York and the Jewish Museum in Berlin.

>>70
Jo Coenen Architects & Urbanists
Jo Coenen graduated in architecture and urban planning from the department of architecture of Eindhoven University of Technology in 1975. Following his studies, he worked until 1979 as a research assistant at the same university and established important professional relationships Europe-wide. Since the start of his academic career, he has taught at various national and international colleges and universities. Has frequently invited to exhibit his work at various Biennials and Triennials.

>>70
Thijs Asselbergs
Is a researcher, connector and strategist. His interest in architecture policy, encouraging young talent and technology are a recurring theme in his work. Early in his career, he was appointed city architect of Haarlem and in the late 1990s, he was one of the reformers of the Dutch building aesthetics. Initiates a program around New Architect which he encourages the younger generation of architects to reflect on the future role of the architect. Has been a professor of Architectural Engineering at TU Delft.

>>70
NL Architects
Is an Amsterdam based office with about 20 international staff. Pieter Bannenberg, Walter van Dijk and Kamiel Klaasse[from the left], educated at the Technical University in Delft, sharing workspace since early nineties, officially opened practice in 1997. They understand architecture as the speculative process of investigating, revealing, reconfiguring the wonderful complexities

>>84

Pancorbo + de Villar + Chacón + Martín Robles

Luis Pancorbo(1969), José de Villar(1976), Carlos Chacón (1977) and Inés Martín Robles(1976) obtained Master degrees of Architect at the School of Architecture of the Technical University of Madrid. Have worked in teams made up of various combinations among the four and focus their professional activity to international open architectural competitions. Have been teaching at various universities and are developing their Ph.D. dissertations. Are also authors of several articles in architectural journals.

>>14

Estudi d'arquitectura Toni Gironès

Toni Gironès graduated from the School of Architecture of the Vallès(ETSAV) - UPC in 1992 and established since 1993 as an independent professional. From 2009 to 2012, he was a member of the Governing Board and professor of the BIArch - Barcelona Ins. of Architecture. Is a member of the board of the "Ciutat Antiga de Vic" as an expert in the field of rehabilitation and heritage. Currently lectures in many professional associations and universities including ETH Zürich, ETSAV and ETSAB. Recently received the first prizes such as FAD Awards, National Spanish Award and Spanish Biennial of Architecture and Urbanism. His works were published in many specialized magazines such as Arquitectura Viva, AV, Domus and exhibited at the SAM(Swiss Architecture Museum) in Basel in 2013 and Catalan Pavilion of Venice Biennale in 2014.

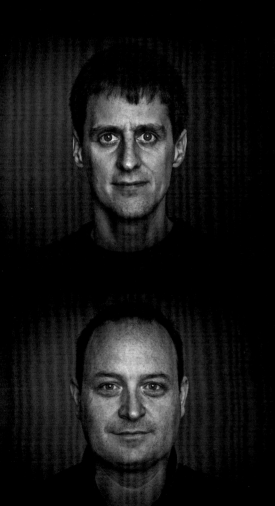

>>26

Savioz Fabrizzi Architectes

Founded in 2004 by the two architects Laurent Savioz[top] and Claude Fabrizzi[bottom], is trying to respond with the best conditions to the needs of the clients by providing all the architectural services from the project to the achievement. Their work is based on the analysis of a site in its natural or built state in order to identify the essentials elements that could enhance, preserve or qualify a site. In this way, the firm enhances the cultural role of the architecture based on the analysis of a function, respectively a program, its place in the history and the culture of a

>>8

JDdVarchitects

Jaco D. de Visser was born in Middelburg, the Netherlands, 1953 and studied at the Academy Arts of Design Artibus(Utrecht section) and Gerrit Rietveld(Amsterdam section). Since then, he studied at the Polytechnic Institute at Amersfoort for 3 years. Established his first own office for Architecture in 1976 and JDdVarchitects in 1988. He aims to stimulate the craftmanship of ideas by the building to and as a catalyst for innovation. Received National Steel Award, Architecture Award at the city of Leusden and Amerstoort, IOC/IAKS Award twice for the quality example of Sports and Leisure.

>>172
R-Zero Arquitectos
Edgar Velasco and Alejandro Zárate graduated from the Faculty of architecture and Urbanism of the Ibero-American University(UIA) and have worked at the Serrano Monjaraz Arquitectos as a Project Coordinator. Edgar Velasco completed a specialty in industrial design at the UIA. Was invited as a architecture critic from the UIA, National Autonomous University of Mexico (UNAM), Western Institute of Technology and Higher Education of Guadalajara(ITESO) and Institute of Technology and Higher Education of Monterrey(ITESM). Alejandro Zárate received a scholarship from the National Fund for Culture and the Arts(FONCA). Before founding R-Zero Arquitectos, he has collaborated with some important studios including Rojkind Arquitectos and was invited as a architecture critic from the UIA and UNAM.

>>138
Office for Strategic Spaces
Angel Borrego Cubero was born in Spain, 1967 and received M. Arch at the Princeton University, USA. After obtaining Ph.D. of Architecture at the Superior Technical School of Architecture of Madrid(ETSAM), he founded Office for Strategic Spaces in 1999. While trained as an architect, he has built a career focused on interdisciplinary topics such as contemporary urban condition and the relationship between private and public space etc. Has taught at Princeton University, Pratt Institute in New York, the University of Alicante, Salamanca University, Keio University in Tokyo, and since 2001 at the ETSAM. Won numerous awards at the national architectural competitions of Madrid and has been nominated for the Mies van der Rohe Award 2014.

Nelson Mota
Graduated from the University of Coimbra in 1998 and received a master's degree in 2006 where he lectured from 2004 to 2009. Was awarded the Távora Prize in 2006 and wrote the book called A Arquitectura do Quotidiano, 2010. Is currently a researcher and guest lecturer at the TU Delft, in the Netherlands. Is a member of the editorial board of the academic journal Footprint and also one of the partners of Comoco Architects.

Douglas Murphy
Studied architecture at the Glasgow School of Art and the Royal College of Art, completing his studies in 2008. As a critic and historian, he is the author of The Architecture of Failure(Zero Books, 2009), on the legacy of 19th century iron and glass architecture, and the forthcoming Last Futures (Verso, 2015), on dreams of technology and nature in the 1960s and 70s. Is also an architecture correspondent for Icon Magazine, and writes regularly for a wide range of publications on architecture and culture.

>>110
Heneghan Peng Architects
Is a design partnership practicing architecture, landscape and urban design. Was founded by Shih-Fu Peng[right] and Róisín Heneghan[left] in New York in 1999 and was relocated to Dublin in 2001.
Have collaborated with many leading designers and engineers on a range of projects which include larger scale urban master plans, bridges, landscapes and buildings. Their major projects include the Giant's Causeway Visitor Center, Central Park Bridges at the London Olympic Park, Grand Egyptian Museum at the Pyramids, a Library and School of Architecture at the University of Greenwich in London and the Refurbishment of the National Gallery of Ireland.

>>124
Naruse Inokuma Architects
Was co-founded by Yuri Naruse[right] and Jun Inokuma[left] in 2007. Is an architectural firm specializing in shared space and creating community-oriented environments with the organizer of the building spaces. Won the International Architecture Awards 2009. Yuri has completed doctorate course in architecture at the Graduate School of Engineering, University of Tokyo. Was a Project Assistant Professor at the University of Tokyo since 2009, and became Assistant Professor in 2010. Jun graduated Graduate School of Engineering, University of Tokyo with a master in architecture in 2004. Has worked at Chiba Manabu Architects until 2006, currently is a Assistant Professor at the Tokyo Metropolitan University.

Heidi Saarinen
Is a London based designer, lecturer at Coventry University and also an artist with current research focused on space and place. Is interested in the peripheral space, in-between and the interaction and collision between architecture, spaces, city, performance and the body. Is currently working on a series of interdisciplinary projects linking architecture, heritage, film and choreography in the urban environment. Is part of several community and c reative groups in London and the UK where she engages in events and projects highlighting awareness of community and architectural conservation in the built environment.

>>154
Neumann/Smith Architecture
Is one of Michigan's largest and most distinguished design firms, honored by over 200 awards from the architectural community and introduced in some international architectural press. The firm has been front and center in Detroit's revitalization, designing innovative spaces that have helped to attract new businesses to the city, moving more than 11,000 people to downtown Detroit. Headquartered in Southfield, Michigan, with a Detroit Design Studio, the firm specializes in architecture, planning, interior design and historic preservation for corporate and municipal offices, mixed-use developments, multi-unit housing, parking structures, commercial and retail centers, and higher education facilities.

>>148
Schmidt Hammer Lassen Architects
Was founded by Morten Schmidt, Bjarne Hammer and John F. Lassen in 1986 in Aarhus, Denmark. Currently has five offices in Aarhus, Copenhagen, London, Shanghai and Singapore. John Lassen was born in 1953 and graduated from the Aarhus School of Architecture in 1983. Has a broad experience as head design architect on most projects that were and are conceived by the firm since its foundation. In particular, he is in charge of the Norwegian projects such as Tjuvholmen Hotel-Apartments in Oslo, the Thor Heyerdahl College of advanced Education in Larvik, and The Northern Light Cathedral in Alta. Is a member of the Association of Danish Architects and Royal Institute of British Architects. Received Eckersberg Medal from the Royal Danish Academy of Fine Arts in 1999.

>>164
1305 Studio
Shen Qiang was born in Shanghai, China in 1978. Since 2001, he has been working in 1305 Studio following Mr. Teng Kun Yen for 13 years. Received several awards such as the Great Indoors Award 2013 Runners Up from FRAME and Golden A'Design Award Winner in Interior Space and Exhibition Design Category.

>>130
Threefold Architects
Jack Hosea, Matthew Driscoll and Renée Searle studied architecture and completed RIBA Part 1, 2 and 3 at the Bartlett School of Architecture, UCL. British architects, Jack Hosea and Matthew Driscoll both began their practical experience at the Michael Hopkins & Partners. They co-established Threefold Architects in 2004. Renée Searle was born in Malaysia and has worked at several London based firms including the Richard Rogers Partnership and Sarah Wigglesworth Architects before joining Threefold Architects in 2009. Three of them have taught at a number of London based and international universities such as the Syracuse University, CEDIM, Central St. Martins and The Bartlett. Some works have been shortlisted in the RIBA Awards, Young Architect of the Year Award and won the 2009 RIBA Award for "The Apprentice Store".

©Franziska Rieder

>>118
KINZO
Was co-founded in 2004 by Karim El-Ishmawi[right], Christopher Middleton[left] and Martin Jacobs[middle] who studied architecture at the Technical University of Berlin(TU Berlin). Karim El-Ishmawi studied architecture at the Illinois Institute of Technology in Chicago after graduation. Is teaching Interior Design at Wentworth Institute of Technology, Boston, USA since 2015. Christopher Middleton has worked at several architectural firms such as Modersohn & Freiesleben in Berlin and Coburn Architecture, Halsted Wells Associates in New York City after graduation. Martin Jacobs participated in several architectural design competitions under the guidance of Kaiser Architects and Axel Schultes Architects and also worked for KCAP Rotterdam on urban design projects.

C3, Issue 2015.5

All Rights Reserved. Authorized translation from the Korean-English language edition published by C3 Publishing Co., Seoul.

©2015大连理工大学出版社
著作权合同登记06-2015年第100号
版权所有·侵权必究

图书在版编目(CIP)数据

办公室景观：汉英对照 / 韩国C3出版公社编；史虹涛等译. — 大连：大连理工大学出版社，2015.9
（C3建筑立场系列丛书）
书名原文：C3 The Changing Landscape of the Office Interior
ISBN 978-7-5685-0134-7

Ⅰ.①办… Ⅱ.①韩… ②史… Ⅲ.①办公室－建筑设计－汉、英 Ⅳ.①TU243

中国版本图书馆CIP数据核字(2015)第222791号

出版发行：大连理工大学出版社
　　　　　（地址：大连市软件园路80号　邮编：116023）
印　　刷：上海锦良印刷厂
幅面尺寸：225mm×300mm
印　　张：11.5
出版时间：2015年9月第1版
印刷时间：2015年9月第1次印刷
出 版 人：金英伟
统　　筹：房　磊
责任编辑：许建宁
封面设计：王志峰
责任校对：王　伟
书　　号：978-7-5685-0134-7
定　　价：228.00元

发　行：0411 84708842
传　真：0411-84701466
E-mail：12282980@qq.com
URL：http://www.dutp.cn